# Equilibrium Models in Economics

# Equilibrium Models in Economics

## *Purposes and Critical Limitations*

Lawrence A. Boland, FRSC

OXFORD
UNIVERSITY PRESS

# OXFORD
UNIVERSITY PRESS

Oxford University Press is a department of the University of Oxford. It furthers
the University's objective of excellence in research, scholarship, and education
by publishing worldwide. Oxford is a registered trade mark of Oxford University
Press in the UK and certain other countries.

Published in the United States of America by Oxford University Press
198 Madison Avenue, New York, NY 10016, United States of America.

Library of Congress Cataloging-in-Publication Data
Names: Boland, Lawrence A., author.
Title: Equilibrium models in economics : purposes and critical limitations /
Lawrence A. Boland. Description: New York : Oxford University Press, 2017. |
Includes bibliographical references and indexes.
Identifiers: LCCN 2016026800| ISBN 9780190274320 (hardcover) |
ISBN 9780190274337 (paperback) Subjects: LCSH: Equilibrium (Economics) |
Econometric models. Classification: LCC HB145 .B65 2016 | DDC 339.5—dc23
LC record available at https://lccn.loc.gov/2016026800

1 3 5 7 9 8 6 4 2

Paperback printed by WebCom, Inc., Canada
Hardback printed by Bridgeport National Bindery, Inc., United States of America

In memory of my former student and long-time friend: the late Dr. Stanley Wong

# CONTENTS

# PREFACE

In the mid-1960s I did my graduate work in a federally financed program that was created to promote and develop what was then high-tech mathematical model building. The main textbooks I used included the large 1958 book, *Linear Programming and Economic Analysis*, by Robert Dorfman, Paul Samuelson and Robert Solow and the small 1957 book, *Three Essays on the State of Economic Science,* by Tjalling Koopmans. Except for a few elective courses and a couple of history of economic thought classes, all of the required courses involved mathematical model building or analysis. The extent of the federal financing was significant since each year of the program provided a generous tax-free three-year fellowship to three new students and it provided a salary for one professor. I say generous because when I took my first job, I had to take a pay cut.

In retrospect, it seems that all of the models we were learning about were equilibrium models – usually Walrasian general equilibrium models. And as such we were learning about existence and uniqueness proofs, stability analysis, and similar issues. We were never required to actually read Léon Walras' 1874 famous book, *Éléments d'économie politique pure, ou théorie de la richesse sociale*; all we were told about Walras was that he engaged in general equilibrium model building. I say 'we were told' to indicate also that we knew nothing of its history. Our two history of thought classes were devoted mostly to learning about economics literature of the eighteenth and nineteenth centuries with little if any mention of Walras. Based on the discussions in the theory classes, and without giving it much thought, I came away with the false impression that Walrasian general equilibrium model building was central to the study of economics from the beginning of the twentieth century.

I raise these strange observations to emphasize a point which Roy Weintraub raised in a 2005 article about the concept of an equilibrium and about the criticisms of equilibrium model building. As I will discuss in Chapter 5, Roy's point is a very important point but those of us in the graduate program in which I was involved would not have understood what Roy was talking about. The point Roy was making was that there are two very different perspectives about equilibrium and equilibrium models – which I will call two cultures in economics. One includes those who learned about the concept of equilibrium before, let us say, 1950, and the other includes those like me and my fellow high-tech PhD students in the 1960s. For us, equilibrium was a property of a mathematical model and we had only a vague idea that it was also supposed to be something about the real world we could see out our windows. For the pre-1950 culture – which was dominated by Marshallian economics – equilibrium was thought to be a claim about what we eventually would or should see in the real world. And the difficulty with all this is that these two cultures both talk about or criticize theoretical states of equilibrium but they are not really talking about the same thing. One of my tasks in this book is to sort these things out so that we can all benefit from each others' criticism.

In my 2014 book on economic model building, I addressed a different schism, the one between today's model builders and those of us who learned decades ago about model building as I did when I was a graduate student. In that book I explained that models and theories were seen as two different things – specifically, we thought that the purpose for a mathematical model was to represent some given economic theory and thereby possibly provide some logical rigour to the theory. When I began working on that book I talked about model building with my colleagues, some young and some old. What I quickly learned was that the young colleagues did not see models as I did. For them the idea of a model was interchangeable with the idea of a theory. By means of a short survey I determined that roughly the year 1980 divided the younger view from my older view. My 2014 book was directed at trying to bridge these two cultures concerning what constitutes a model in economics. Interestingly, for that schism between the older and younger views of the relationship between theories and models, I was a member of the older side. But in the present book, which will be addressing the schism that Roy identified concerning the concept of an equilibrium, thanks to my training in the high-tech graduate program, I became a trained member of the younger side of Roy's schism.

Ironically, despite the best efforts of my graduate instructors, once I began teaching the ubiquitous Economics 101 class I realized how useless my graduate training was when it came to understanding the real world so that I could teach about it. Early on I deviated from my training and began teaching my students about equilibrium as something about the real world much like the older side of Roy's schism did. In the process I discovered Joan Robinson and read many of her criticisms of the work of the newer side of Roy's schism. When I later got to teach the fourth-year advanced microeconomics theory seminar I started looking at some interesting articles that were about how the concept of an equilibrium was problematic in economic explanations. As it turned out, these critical articles were all addressing problems with formal equilibrium models. Moreover, thanks to many of my critical students I learned a lot about economics and economic model building by later teaching an advanced micro theory seminar and then even more when I began teaching a graduate micro class. I think what I learned in those classes I should have learned in graduate school. As a result, I have decided that this book will be about what I learned with my students about equilibrium concepts and equilibrium models.

This book will be addressing recognized problems with equilibrium models particularly from the perspective of standard economics textbooks that use equilibrium models as a basis for explaining prices or forming economic policies and especially in teaching beginning students the virtues of the competitive market. Of particular concern will be how economics textbooks almost always fail to recognize any problems with equilibrium models even though these problems fundamentally distort realistic economic explanations. So, as I go along and whenever possible, I will try to point out ideas and criticisms that are relevant today but have their origin in the ideas published by economic model builders decades ago. While my main interest is in what we teach students, eventually what will be considered here might also enable us to explain why most governments' policy makers are failing to provide effective help dealing with real world economies. After all, most governmental economic policy makers likely were once students in an Economics 101 class.

In 1986 I published a methodology book that was also about what I learned teaching both advanced and graduate microeconomics theory classes. That book proposed to offer a new methodological perspective for addressing some fundamental problems with common microeconomic models. Unfortunately, almost all of the problems I discussed there still seem to persist in microeconomic model building today, particularly with those that rely on using the analytical properties of equilibrium states. While in this book I will be dropping most of the methodological concerns of that book, I will be returning to many of the theoretical problems I discussed then, but this time by focusing instead on recognized problems involved in building equilibrium models. While methodology will play a much lesser role than it did in the 1986 book, it will be addressed briefly in Chapter 6 and a bit more in Part III, where I discuss how common methodological presumptions constrain any attempts to solve the problems I discussed in Parts I and II.

I have written this book for readers interested in learning about the main tool economists use to help understand the economy. Such readers include undergraduate and graduate students, of course. But I also hope readers who may not have taken the proverbial Economics 101 – or, if they did, do not remember much from that class – will still find this book useful. For these readers I will occasionally add footnotes to help with the usual economists' jargon that one would have learned in that class. And most important, it is this group of readers in which we will find people employed as governmental advisors and policy makers – in particular, people who should be asking economists about the assumptions that were used to reach the advice they are giving advisors and policy makers.

# ACKNOWLEDGEMENTS

I have received a considerable amount of help in the form of criticisms of early versions of chapters for this book. For this help I wish to thank my former students Senyo Adjibolosoo and David Hammes as well as colleagues Brian Krauth, John Knowles, Ken Kasa and Luba Peterson and friends Pedro Garcia Duarte and Mark Donnelly. Also, I thank Duncan Foley for helping me with Chapter 12 and Kenneth Arrow and George Richardson for answering my questions about their articles discussed in Part I.

Also note that I have made use of parts of several chapters from my 1986 *Methodology for a New Microeconomics: The Critical Foundations* that was published by Allen and Unwin. None of those chapters are reproduced here as in each case the material I have used has been heavily revised as well as updated. Any reader interested in that 1986 book can now obtain a 2014 Routledge Revivals edition published by Taylor and Francis publisher.

# Equilibrium Models in Economics

# Prologue

## *Problems with modelling equilibrium attainment*

The idea of a state of equilibrium pervades economics research. For a definition of an equilibrium one can easily find one with a Google search and see something like this found in the 2016 edition of *The American Heritage Dictionary of the English Language*: 'e·qui·lib·ri·um: A condition in which all acting influences are canceled by others, resulting in a stable, balanced, or unchanging system'. If one also looked for how a well-known economist might view the notion of an equilibrium in economics, no better example to be found would be the view of Frank Hahn: 'Whenever economics is used or thought about, equilibrium is a central organising idea. Chancellors devise budgets to establish some desirable equilibrium and alter exchange rates to correct "fundamental disequilibria"' [1973, p. 1].

Critical analysis of equilibrium models is not a new topic. Of particular interest for any consideration of claimed limits to equilibrium-based explanations are three separate and different challenges presented in, ironically, the same year – namely 1959. These challenges focused on equilibrium models of the market that were very common in economics texts then and are still common today, particularly in undergraduate textbooks. One of the three 1959 articles was by Kenneth Arrow who simply pointed out that in the usual *equilibrium market model* (such as that illustrated in Figure P.1) it is not enough to explain a market equilibrium price ($P_e$) as the one at an intersection between the demand curve and the supply curve. He insisted that one must also explain *in the model* why that equilibrium price would be achieved. As will be explained in Chapter 2, the problem is that even to discuss a market-determined *equilibrium* price one needs to recognize that a stable arrangement of the market's

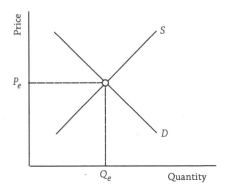

**Figure P.1.** Simple market equilibrium

demand and supply curves is required *within* one's model. But, as Arrow also recognized, a stable arrangement is not enough because with the usual textbook models we are never told how any market participant, say the supplier, knows the market's demand curve or at least knows when to lower the price and by how much. Similarly, how does the demander know when to bid up the price or know by how much? Textbooks just rely on some vague form of allowance of a sufficient amount of time but never say how much time this would take. As to how the price was determined within the model, Arrow suggested one possible solution for this problem of adequate explanation was to recognize that the usual textbook discussion of an *imperfect competitor* does involve at least a supplier setting the price[1] – but, of course, this would require recognizing the supplier's knowledge, and learning or at least identifying the available information needed to make such a decision.

Ironically, in 1959 Robert Clower published an equilibrium model about a different problem but one that in effect directly addressed Arrow's suggestion.[2] Clower's model was of an 'ignorant' monopolist for which allowance is made for the obvious fact that the monopolist could not possibly know the whole market demand curve it faces but instead would have to make assumptions about it. For Clower, the ignorant monopolist would at least have to

1. For those unfamiliar with economics jargon, perfect competition refers to a type of market in which no individual has a significant role or effect on the determination of the equilibrium price and imperfect competition means that individuals can affect the price. Textbooks distinguish between these two types of competition by claiming that perfect competition will exist whenever there are too many participants for any one to have an effect and competition is imperfect whenever the number or buyers or sellers is small enough that every participant can have an effect because any change in their behaviour affects either the market's demand or the market's supply. This distinction plays a big role in textbooks' definition of markets and market behaviour.
2. I say 'ironically' because I asked Arrow (in January 2014) if he was aware of Clower's article in 1959 and he said he was not.

make assumptions about the shape and position of the market's demand curve. Based on those assumptions, Clower's monopolist would send a chosen supply quantity (presumed to be a profit-maximizing quantity) to the market and wait to see what market-clearing price is obtained.[3] If the assumptions about the nature of the demand curve are true – such that the *implied* marginal[4] revenue for the supplied quantity would be equal to the marginal cost for that quantity – there would be no problem.[5] But there is no reason to think the monopolistic supplier has the required knowledge to assure that the assumptions made about the market's demand curve are true. As will be explained in Chapter 3, the result is a model in which an apparent equilibrium price may be reached but it is one at which the monopolist is not actually maximizing profit even though the monopolist is erroneously thinking it is.

And again ironically in 1959, George Richardson presented a model of the competitive market where it would seem that the only way to *guarantee* the attainment of a market's equilibrium price is to introduce some form of collusion.[6] As will be explained in Chapter 4, if Richardson is right, then this necessity would obviously fundamentally challenge what is commonly taught in 'Economics 101' class.[7]

Richardson recognized that Friedrich Hayek in a 1937 article had already raised concerns about the knowledge requirements for the achievement of a market's equilibrium. It turns out this was preceded by a 1933 lecture in which Hayek suggested there was an even more fundamental problem concerning the information available to an investor.[8] Hayek was concerned that it is too

---

3. For those readers who have never taken an economics class or do not remember much from of what they heard in their economics class, maximizing profit just means maximizing sales revenue net of production costs.

4. For those not remembering economics jargon, the word 'marginal' is just jargon for the following idea. If one is deciding about increasing the amount to produce of some commodity, marginal refers to the consequences of an increase of one unit of that commodity. In the case of revenue, it is how much more revenue is obtained if one sells one more unit of the commodity in question. Marginal cost would then similarly be about the change in total production cost if one more unit is produced.

5. Again, for those not remembering their economics jargon, this equality is just a matter of whether profit (sales revenue net of production costs) is maximized. For it to be at maximum, calculus textbooks tell us there must be an equality of marginal revenue and marginal cost. If they are not equal, then a gain in net revenue is possible by producing either more or less depending on whether the difference between them is positive or negative.

6. I also asked Arrow if he was aware of Richardson's article when he wrote his own article and again he said he was not. So then I asked Richardson (through his son Graham) if he was aware of either of the articles by Arrow or Clower and he said he was not.

7. This is the jargon name given to the usual beginning economic principles class offered in university and college programs.

8. Hayek delivered this lecture on December 7, 1933, in the *Sozialökonomisk Samfund* in Copenhagen and which was first published (in German) in the *Nationalökonomisk*

easy for governments, through a policy (perhaps by manipulating the interest rate in the money market), to actually give false information unintentionally to investors in the current capital equipment market and thereby cause disequilibria in future product markets. While Hayek may have raised the issue as a criticism of government intervention when facing the problems of the Great Depression in the 1930s, the issue he raised does recognize the limitation of models of the market that do not recognize how decision-makers in the model know what they need to know to assure the achievement of equilibria when investment decisions are involved. This is particularly the problem when a firm is placing an order for capital equipment which of necessity takes time to produce. Perhaps by the time the equipment is delivered to the firm the market does not resemble one that the investor might believe would be there as suggested by what the change in government policy promised.[9]

## P.1.  A GENERAL ECONOMIC EQUILIBRIUM AS A NECESSARY SOCIAL OPTIMUM

In the Richard T. Ely Lecture to the American Economics Association's 1994 conference, Arrow [1994, p. 4] told us about what he called the prototypical economic model:

> The prototypical economic model ... is general competitive equilibrium. Individuals and firms take prices as given. Individuals choose consumption demands and offers of labor and other assets, subject to a condition that receipts cover expenditures. Firms choose inputs and outputs subject to the condition that the outputs be producible given the inputs. How they make these choices depends on many factors: tastes, attitudes toward risk, expectations of the future. But, it is held, these factors are *individual*.

As Adam Smith's eighteenth-century view of the world would have us believe, we should never depend on authorities such as the Church or the state since the 'best of all possible worlds' will be achieved when everyone is independently and autonomously pursuing self-interest and nobody is inhibited in that pursuit except by the limits imposed by Nature. With this in mind, let us examine the world where everything about the economy is a matter

*Tidsskrift*, vol. 73, no. 3, 1935, and later (in French) in the *Revue de Science Economique*, Liège, October 1935.

9. Perhaps it should be noted that the going interest rate in a state of long-run or general equilibrium is sometimes seen to indicate the equilibrium growth rate of the economy – see John von Neuman [1937/45].

of individual choice except natural givens. Of particular interest, we need to know why economists would ever claim that it is the 'best of all possible worlds'. The world they are talking about is one in which: (1) everyone is free to enter or exit from any market (there are no non-natural constraints), (2) there is a market for everything in the production process (which implies the supply of all inputs are to some degree variable) and (3) when an equilibrium is reached, all participants are independent and autonomous optimizers (maximizers or minimizers) which means every participant in any market must be individually optimizing and thus simultaneously every market must be in equilibrium – in other words, this means for the economy as a whole there must be a general equilibrium. If anyone were not optimizing, then, necessarily, that individual has an incentive to change his or her behaviour (i.e., his or her demand or supply of some good or service).[10] For this world, any general equilibrium is therefore a social optimum since everyone is individually optimizing.

Before going much further talking about equilibrium models – and for the benefit of any readers who do not remember their Economics 101 class[11] – let me briefly and carefully review what students would usually learn in beginning textbooks. To do this – and for those remembering their Economics 101, so we are all on the same page – let us examine a very simple elementary general equilibrium model of a very small economy.

A keystone idea of any equilibrium model that might be used to explain an economy is the distinction between those variables represented in the model which the model is designed to explain, and those variables in the model that are not to be explained *within* or by the model but are still claimed to be relevant nevertheless. Most economics textbooks discuss this distinction and call the variables to be explained 'endogenous' variables and those relevant but not to be explained are called 'exogenous' variables.[12] For this small elementary model, let us consider a simple textbook's world consisting of just two commodities ($X$ and $Y$), two inputs used in the production of those commodities ($L$ and $K$),[13] and two individuals ($A$ and $B$). The equilibrium model of

---

10. Usually, textbooks will say reaching an equilibrium state means reaching a state in which variables such as prices or demand and supply quantities are no longer changing.

11. Or who may not have benefited from taking such a class.

12. Here I am talking about variables. There are other possible ingredients which are called parameters or coefficients that are the resulting properties of how a model builder chooses to represent the behavioural equations in the model. Let me leave them alone for now as here I will focus only on observable variables recognized within the model. Parameters and coefficients are not usually observable.

13. For those who have never taken Economics 101, $L$ represents labour and $K$ represents physical capital like machines.

any such small economic system claims to describe the determination of the following *endogenous* variables:[14]

| | | |
|---|---|---|
| Prices: | $P_X, P_Y, P_L, P_K$ | |
| Quantities: | $X, Y$ | (total demands and supplies of the commodities) |
| | $X_A, Y_A, X_B, Y_B$ | (individual consumer demands) |
| | $L_X, L_Y, K_X, K_Y$ | (individual producer demands for inputs) |
| | $L_A, L_B, K_A, K_B$ | (individual supplies made available) |
| Industry sizes: | $N_X, N_Y$ | (the number of respective firms) |
| Incomes: | $I_A, I_B$ | |
| Utility levels:[15] | $U_A, U_B$ | |
| Transformation rates: | MRS | (marginal rate of substitution between $X$ and $Y$)[16] |
| | MRTS | (marginal rate of technical substitution between $L$ and $K$)[17] |

Note that none of these are necessarily exogenous variables.[18]

For those readers unfamiliar with economists' jargon, when textbooks say 'determine' they usually mean 'explain' in the sense that for the given values or states of the exogenous variables and the model's behavioural assumptions

14. The subscripts indicate to whom or what the variable refers to.

15. For those unfamiliar with today's textbooks, 'utility' is often referred to as a measure of a consumer's satisfaction – and as such being 'more' or 'less'. But this unfortunately suggests that we can measure our utility or assign any level with a cardinal number (e.g., 22 'utils' vs. 20 'utils') as one would assign 'degrees' to a temperature. But, the terms 'more' and 'less' only refer to a relative measurement and nothing more – no cardinal number intended. And for this reason, textbooks that wish to make clear that they are not referring to maximizing cardinal utility usually refer to someone maximizing satisfaction – as if satisfaction could not be assigned a cardinal number. All that is intended is that it is still an exact level of satisfaction being sensed *by the individual* even though he or she cannot put a number on it – and for that reason it is just a matter of more or less. Most importantly, textbooks always presume that all individuals simply know when they are better off.

16. Technically, textbooks at this point ignore the matter of not being able to assign cardinal numbers and thereby let marginal rates of substitution refer to ratios of respective 'marginal utilities' (the ratios of the additional 'amounts' of utility or satisfaction from consuming *one more unit* of each of the two goods being consumed).

17. This is the name given to the ratio of 'marginal productivities' (the ratio that the additional amounts of the output would change from using one more unit of each of the two inputs). The problem of measurability is obviously not an issue here.

18. It should be allowed that when textbooks distinguish between short-run and long-run equilibria, they might treat an endogenous variable whose equilibrium value is determined only in a long-run equilibrium model as a static 'given' in the determination of the short-run equilibrium. Typically, the available amount of physical capital

relating all the variables and givens, one can show that each of the above 26 variables have particular values (and this is usually accomplished by solving for the values of the set of the endogenous variables using a model consisting of a system of 'simultaneous' equations with which the endogenous and exogenous variables are represented).

If the world implied by the model is in a state of general equilibrium, how could anyone claim that it is not an optimum world, that is, claim that it is not a 'best of all possible worlds' since it is also assumed that in the model *everyone* is personally maximizing? If we were to claim that it is not, then we would be saying that we know better than the market participants themselves – that is, we would have to claim that at least one individual is not maximizing even though he or she may think otherwise. Unless we have access to some extra variables which are not already recognized and represented in this model's general equilibrium world, there is no way for us to know more than any individual participant. And such extra variables cannot be among the endogenous variables since the latter are already determined by the interaction of all of the model's individuals. Thus, the extra variables must be exogenous. If we were participants in the market, we would be in a position to gain by our privileged access. Such a potential gain would mean that the model's market was not actually in a state of equilibrium. Even if one has to be outside the world created by the model to be able to claim that a given general equilibrium is not an optimum, the given equilibrium may still be the best of all 'possible' worlds – that is, the equilibrium may be possible for the model's individuals acting without *outside* help.

The question of the social optimality in any given model of general equilibrium also connects the optimum for the model's whole economy with the numerous personal optima of all of the model's independent and autonomous individuals separately. For example, if all of the model's individuals are maximizing, the (linear) sum of their maxima is itself a maximum. Whenever each individual is at a point for which being at any other point means non-optimality, the aggregate of all individuals' choices will also be an optimum.[19] In this case, a general equilibrium in this model's world is a social welfare optimum, in the sense that should any individual deviate from his or her personal optimum, the aggregate welfare will be reduced. And again, for us to say that it is not the 'global' optimum requires us to have an outside perspective that is precluded by definition of the model's world of autonomous individuals.

All this is quite consistent with the idea of a market equilibrium presented in textbooks, such as the equilibrium illustrated in Figure P.1. In the

---

is such a variable because it is usually assumed that such capital takes more time to produce than those variables being explained in the short-run model.

19. See Koopmans [1957, pp. 50–1].

textbooks' so-called neoclassical theory of prices,[20] the demand curve (D) is the locus of all price and demand quantity points for which at each point every demander would be maximizing his or her utility. Similarly, the supply curve (S) is the locus of all price and supply quantity points for which at each point the profit of each and every supplier is being maximized. When the price is the one at which the market clears (i.e., when demand equals supply), the price is also the one at which each individual (by maximizing) is choosing the correct quantity to demand or to supply. At market-clearing prices, the market's aggregate supply and aggregate demand are also equal, even though no individual has to calculate or even be aware of such aggregates.

## P.2. MAXIMIZATION IS THE ONLY BEHAVIOURAL ASSUMPTION IN NEOCLASSICAL EQUILIBRIUM MODELS

It is important to recognize – although rarely it is recognized – that the only behavioural assumption in neoclassical equilibrium models (including the simple equilibrium model above) is that every individual recognized in the model is a maximizer – maximizing utility or satisfaction in the case of the demanders of consumption goods and profit in the case of the producers who supply them. By saying it is the only behavioural assumption, I am particularly saying that the common additional textbooks' assumption of a state of equilibrium is redundant at best, misleading at worse. Adding an assumption of a state of equilibrium is misleading to the extent that its inclusion falsely suggests that such an assumption is necessary. The redundancy by now should be apparent. If everyone is maximizing as usually assumed, then there is no behavioural reason for anyone to change their choices concerning endogenous variables (such as the quantity demanded or supplied) until there is a change in one or more exogenous givens. And obviously, the prime attribute of any state of equilibrium is the absence of change.

It is also important to keep in mind that this singular behavioural assumption is also double edged. It can be used to explain the absence of change and it can be used to explain the reason for change. Specifically, being in a position where an individual is not maximizing is the primary reason for that individual to change his or her choice. Whenever there is excess demand – that is, whenever the demand for a good exceeds what is being supplied in the market – some demanders are not able to maximize their utility because their utility-maximizing demand quantity is not available. As will be discussed in Chapter 2, any failure to maximize plays a significant role in explaining the equilibrium process.

---

20. This is just the usual name given by historians of economic thought to what is taught in Economics 101. Classical economics usually refers to late eighteenth century economics and neoclassical to the economics developed in the late nineteenth century. Both include what I identified with the three ideas (1) to (3) at the top of this section.

It should also be noted that this singular behavioural assumption plays a significant role when we consider what it means to attain a model's general equilibrium. That is, for any particular market to truly be in equilibrium, all other markets in the model's whole economy would have to be in equilibrium. If they are not all in equilibrium it would mean that at least one participant in some market is not successfully maximizing. For example, consider the world of the simple model described above in Section P.1. For it to include an individual who is maximizing with respect to the purchase of good $X$, the price for the other good, $Y$, must be an equilibrium price. If it is not, then just what price is it? If it is the equilibrium price for $Y$, in principle the optimum choice for $X$ already implies the optimality of the demand for $Y$ as well.[21]

## P.3.  ON THE ROAD TO DYNAMIC STOCHASTIC GENERAL EQUILIBRIUM MODELS

While it would be tempting to discuss all of this in the context of the history of economic thought, that will not be the focus of the discussions in this book. Instead, I will be adopting the same posture raised by Takashi Negishi who said, 'What is important is not whether a particular interpretation of a past theory is correct, but whether it is useful in developing a new theory in the present' [1985, p. 2]. That is, whenever I raise historical observations it is not just to emphasize that the issues to be discussed have been around for a long time but also that it is time to address them because I think they have relevance for models being built in today's economics.

Today, equilibrium model building – particularly using the usual textbooks' so-called Walrasian general equilibrium model[22] – is central to virtually all aspects of economic research. The most prominent example is the now commonly used Dynamic Stochastic General Equilibrium (DSGE) model which is based on the work of Arrow and Gerard Debreu in the 1950s. Today, versions of the DSGE model dominate empirical macroeconomics-based model building even though – as I explained in my 2014 book – there are several recognized problems

21. Given that there are only two goods in this simple model and that an individual consumer's optimizing choice involves deciding how to allocate his or her budget between these two goods (such that buying more of $X$ means demanding less of $Y$), an equilibrium cannot exist in one market without there being an equilibrium in the other market – see Hicks [1939/46, pp. 66–7].

22. Historically, what textbooks call the Walrasian general equilibrium model is a system of simultaneous equations in which each equation represents a necessary calculus condition for optimization, and thus together the equations represent the necessary conditions for all markets to be clearing (i.e., to be in equilibrium). The origin of such a concept of a formal general equilibrium model is usually thought to be due to Walras [1874] – hence the name – but the version textbooks refer to is really the simplified version created by Gustav Cassel [1918/23].

with basing macroeconomics on Walrasian general equilibrium models when it comes to obvious topics such as the role of money in the economy.

One problem, recognized long before the creation of the DSGE model, concerned the process of attaining a model's state of general equilibrium – or for that matter, for attaining the textbooks' model's long-run equilibrium. Specifically, what is assumed in the model about the knowledge obtained by the market participants that will assure the attainment of the equilibrium? As I noted in the brief discussion of Richardson's article, this was noted eighty years ago by Hayek and further explained in Richardson's article that will be discussed in Chapter 4.

## P.4. OUTLINE OF THIS BOOK

This book is not intended to be a textbook on building equilibrium models – although I would hope it might be of help to students and teachers of fourth-year seminars about economic theory and model building as well as of early graduate classes that involve those subjects. However, as I noted in the Preface, I have kept in mind that some readers may not be students or teachers but readers interested in figuring out what economists do and what economics students learn. Of course, these readers may not be familiar with the economists' jargon and so – as I have been doing already – I will always try to explain such matters in some of the footnotes.

The problems with equilibrium models will be explored throughout this book starting in Part I with a general discussion in Chapter 1 about using equilibrium models to explain economic variables, which will be followed by three chapters each discussing one of the interesting articles I have my students read concerning some recognized fundamental problems with all equilibrium models. These are the three articles by Arrow, Clower and Richardson that I briefly discussed at the beginning of this Prologue.

Part II will examine the alleged purposes and limitations of all forms of equilibrium models, and finally, Part III will critically examine three key ways that model builders can attempt to overcome the limitations of the textbooks' equilibrium explanations. The first way examines theorizing about price dynamics as suggested by Arrow. The second way involves attempts to avoid beginning with the assumption of a pre-existing general equilibrium by instead focusing on non-clearing markets. This second way will be an attempt to reconstruct the disequilibrium basis of the anti-long-run equilibrium economics of John Maynard Keynes. And the third way involves addressing the need to include explicit assumptions about learning and the knowledge necessary to assure the attainment of a state of general (or partial) equilibrium – that is, it will address a significant aspect of equilibrium model building that is missing from economics textbooks and teaching.

# PART ONE

# The purpose and problems
# for equilibrium models

# CHAPTER 1

༦ᘉༀ

# Equilibrium models and explanation

et us begin by considering a textbook's view of equilibrium. For this view, consider my friends' and my colleague's textbook, *Microeconomics: Theory and Applications*, by B. Curtis Eaton, Diane Eaton and Douglas Allen in which they explain [2012, p. 19]:

> The economist's aim is to explain or predict the social state that will arise from the choices of the individual economic participants. The method used to make these predictions is the method of equilibrium. At the heart of this method is the concept of an equilibrium. It can be defined this way: *an equilibrium consists of a set of choices for the individuals and a corresponding social state such that no individual can make himself or herself better off by making some other choice.*

Of course, today it is widely recognized that equilibrium models are commonly used to explain variables of interest. As noted in the Prologue, whether one begins with the behavioural assumption that everyone is maximizing and hence equilibrium is assured, or one just jumps to presume the existence of an established equilibrium state, does not matter much for most model builders or textbook writers. With this in mind, this chapter is about how a presumed state of equilibrium[1] is used as a method of explanation.

The concept of an equilibrium has been central in economics for over 200 years, that is, easily since the time of Adam Smith.[2] For Smith and many

---

1. Usually in the case of formal equilibrium models, a state of equilibrium is obtained just by assuring that there is a solution to the set of simultaneous equations that constitute the model.

2. For an excellent history of the development of the notion of an equilibrium in economics, see Milgate [1987]. For an equally excellent discussion of all the various ideas about an equilibrium, see Samuels [1997].

of his followers the concept has often been used to explain away supposed evil human tendencies such as 'greed'. As children, most of us were taught that greed is a social evil. Nevertheless, we must not despair since, if we follow Smith, it can be shown that such a greediness constrained by the eventual state of equilibrium can be seen as a 'virtue' rather than a 'vice'.[3] That is, by reaching a state of competitive equilibrium, greed will actually be shown to have led to the good of everyone. In such a state each individual is personally optimizing, given his or her respective resources, and there is no way any self-interested individual could ever get ahead except by being greedy.

Teachers of Economics 101[4] do not usually put the idea of competitive market equilibrium in such stark terms as vice and virtue. This is partly because since the late nineteenth century economic theories and models have become the main 'scientific' foundation for explaining and understanding the economy. And the main tool has been the idea of a competitive general equilibrium which is nevertheless just an elaboration of Smith's perspective.

## 1.1. EQUILIBRIUM AND EXPLANATION: ELEMENTARY CONSIDERATIONS

Today the concept of a state of equilibrium is usually embraced for reasons other than its role in Smith's social philosophy of private goods and social evils.[5] The reasons are to be found in Alfred Marshall's self-conscious theory of 'scientific explanation' which today is called 'comparative statics'. Marshall said, comparative statics 'is the only method by which science has ever made any great progress in dealing with complex and changeful matter, whether in the physical or moral world' [1920/64, p. 315, fn 1]. Marshall's comparative statics explains things in a very special way and is based on using equilibrium models in which endogenous and exogenous variables play a central role.[6] The quotation above is from Book V of Marshall's famous textbook, *Principles of Economics*, first published in the late-nineteenth century. Even today most

---

3. Perhaps it should also be recognized that Bernard Mandeville in his seventeenth century *Fable of the Bees* saw this juxtaposition long before Smith.

4. Again, this is just jargon referring to the ubiquitous beginning economic principles class but it is not likely the exact name used in every economics department.

5. In this and the next two sections I will be applying some of the arguments I presented in Boland [1986].

6. As I noted in the Prologue for the benefit of readers unfamiliar with economists' jargon, the variables that we want to explain within our model are the endogenous variables and their explanation is always conditional, that is, endogenous variables depend on the values of exogenous variables which are not explained within the model.

microeconomics principles textbooks just present a version of Marshall's Book V.[7] In this section I will discuss the elements of Marshall's equilibrium-based explanations that are taught to students.[8] Unfortunately, many of the textbooks' renditions of these elements are inadequate at best and misleading at worst.

### 1.1.1. Marshall's two 'Principles' of explanation

Before going further, it is important to recognize that Marshall's model of explanation employs only two 'Principles'. The one all textbooks recognize is the singular behavioural assumption of maximization which Marshall called the 'Principle of Substitution'. For Marshall substitution meant, for example, that the individual consumer would be seen to be substituting one bundle of goods (specific amounts of, say, $X$ and $Y$) for another bundle (with different amounts of $X$ and $Y$) because it will provide more satisfaction. The other principle is equally important but never mentioned in textbooks. It is his 'Principle of Continuity', which simply says that one cannot be claimed to be making a maximizing choice unless there is a continuous range of options to choose between.[9] Without such a range, there would not be a choice to make that can be explained.[10]

His Principle of Continuity is the basis for distinguishing his short run from his long run. Specifically, for the firm he defines the short run by saying that the short run is too short for the firm to change its amount of available physical capital $K$ (e.g., machinery), hence the choice of $K$ cannot be explained with a short-run equilibrium model. However, his long run is long enough for $K$ to be changed so that a profit maximizing choice can be made and thereby explained.

And finally, given that in the nineteenth century when he was writing his book, most economies were dominated by agriculture, he defines the short-run to be a length of time corresponding to what it would take to plant seeds and eventually harvest the resulting crop. In his Book V he explicitly says the short run amounts to 'a few months or a year' [1920/64, p. 314]. His long run might last a year or more depending on what is required for the production of the machinery.[11]

---

7. What some Victorian authors such as Marshall called Books today would just be called Parts.

8. The following discussion about Marshall's mode of explanation is not an exercise in the history of economic thought because what is taught today in Economics 101 is nothing more than Book V of his *Principles of Economics* book.

9. That is, it is possible to have more or less than the amount chosen.

10. For a fuller discussion of Marshall's two principles and the role of time in his mode of explanation, see Boland [1992, chs. 2 and 3].

11. Marshall [1920/64, p. 315] actually identified his long period as 'several years' which is enough time for capital to be varied either in quantity or in kind needed to

## 1.1.2.  Long-run vs. short-run equilibria and the role of time

One of the most misunderstood of Marshall's tools involves the conception of time – specifically his temporal distinction between short-run and long-run equilibria.[12] Students are led to believe that these refer to time going forward such that the short-run equilibrium occurs after a short period of time in the future and the long-run equilibrium occurs even further in the future. But this is completely wrong, as I will explain.

Marshall's distinction between short and long runs arises when he explains the firm's choice of the observed amount of labour being used today. Specifically, he asks us to see this choice as a result of a process long enough for the amount to be changed. Since it is relatively easy to change the amount of labour to use, there would be enough time in the short period to choose an amount $L$ for which profit *has been* maximized. That is, his explanations amount to looking back in time. His explanation is that the firm is recognized to have had a sufficient period of time to make the optimizing choice of $L$. Similarly, his explanation of the firm's choice of the profit maximizing amount of physical capital $K$ presumes that the firm is recognized to have had a sufficient period of time to make that choice and that would require a look further back in time than the explanation of $L$ would require. And, given this, it is also important to recognize that every long-run equilibrium is thus a short-run equilibrium since looking backward, by the time the long-run equilibrium has been reached there would have been enough time to reach the short-run.

So, with all this in mind, Marshall's mode of explanation is simply to characterize the firm's observed current amount of labour, $L$, to be that corresponding to the firm in a state of short-run equilibrium and similarly the amount of capital observed in use to correspond to the firm being in a state of long-run equilibrium. These equilibria exist today, the day of observation, not some day in the short-run future or the respective long-run future. This means that the Marshallian mode of explanation involves recognizing a transpired period of time. Nevertheless, Marshallian equilibrium analysis is often criticized for being timeless.[13]

move to produce a different good for a different industry. Even in his long period, he assumes technology is fixed and does not change for a generation (it requires a generation since a change in technology may have to wait until the children take over the farm as they may be more willing to use more modern technology). Evidently, his perspective really was that of an agrarian economy.

12. Marshall actually distinguished three different 'runs' or, in his words, 'periods'. In addition to his short period and his long period he also recognized a 'market period', which is when the price is determined. This idea is that at a farmer's market the supply is delivered in the morning and it does not change during the day so that only the price can change to assure the market clears by the end of the day. Textbooks today do not usually discuss this market period.

13. Usually, this is because the critics see an equilibrium to be a static or unchanging state. Such critics will be discussed in Chapter 5.

In the Prologue a simple equilibrium model was discussed in terms of the observable endogenous and exogenous variables. But, of course, from the perspective of formal mathematics the model requires assumptions going beyond the observables. The model builder does this by specifying functional relationships between the variables. In some cases the variables will be related by means of a linear mathematical function, while in other cases they will be related by something nonlinear. Maximization usually requires nonlinear relationships. The exact mathematical formalities of all this will not be of interest here as the main objects of interest are just the resulting values for the observed endogenous variables given the observed values of the exogenous variables. Should the model builder assume that the relationships are empirically correct, then the model will be expected to yield correct values of all of the observed endogenous variables given the observed exogenous variables.

By building such an equilibrium model, economists would say that the resulting equilibrium values of the endogenous variables have been explained. The explanation amounts to a logical argument formed of the assumptions that relate the model's variables to the values of the observed exogenous variables. It also amounts to a 'causal' explanation at the level of saying the *set* of exogenous variables *causes* the *set* of the endogenous variables. In this regard, it is tempting to see any singular exogenous variable as being the cause of the equilibrium values of the endogenous variables. As I will explain later in Chapter 5, the idea of a single exogenous variable being a cause of the equilibrium values is misleading since all of the many other exogenous variables are also playing a role.

### 1.1.3. Comparative statics analysis as thought experiment

One question that will nevertheless be asked of an equilibrium model is about the role played by any given exogenous variable in the determination of the equilibrium values of the endogenous variables. For this, economists employ what they call *comparative statics analysis*. In comparative statics analysis two different sets of equilibrium values for the endogenous variables, representing two different states of equilibrium, are compared. The two equilibrium states are distinguished *only* by considering two different values for a single[14] exogenous variable recognized in the equilibrium model.

The careful application of the method of comparative statics analysis has been the primary basis of almost all of our understanding of the economy when equilibrium models are the basis for our explanations. Even the so-called

14. Of course, it is possible to consider more than one exogenous variable but it would not be clear what one can conclude from that. The purpose of changing just one

multiplier in old textbook Keynesian macroeconomic analysis is based on this method. For example, the macroeconomics model builder might ask what would be the effect on the equilibrium level of national income that would result from increasing the level of investment *in the model* as a thought experiment. The effect on national income by say a ten percent increase in investment can be determined by using the model by solving for two sets of values for the endogenous variables including the variable representing national income that would result from the two different values of the exogenous investment variable. In other words, by means of comparative statics analysis, equilibrium models can be used to perform a thought experiment in order to understand the causal role of any exogenous variable in the explanation of that model's endogenous variables.

## 1.2. EQUILIBRIUM IMPLIES RECOGNITION OF DISEQUILIBRIUM DYNAMICS

In his 1994 Richard T. Ely Lecture, Arrow also notes [p. 4]:

> Even if we accept this entire [general equilibrium] story, there is still one element not individual: namely, the prices faced by the firms and individuals. What individual has chosen prices? In the formal theory, at least, no one. They are determined on (not by) social institutions known as markets, which equate supply and demand.

A main aspect of Marshall's mode of explanation using equilibrium models was left out of my discussion above. That is his use of what is called the assumption of *ceteris paribus*. As I have already noted, the equilibrium values of the model's endogenous variables will not change so long as the model's exogenous variables do not change – that being the main point of distinguishing between these two types of variable. And since the use of either short-run or long-run equilibrium models involves the passage of time, and for his explanation to be logically adequate, as time passes during the run none of the exogenous variables can change. For this reason the assumption of *ceteris paribus* must be included in the model's explanation. This is because the explanation of those endogenous variables requires *ceteris paribus* for those variables to be able to reach their equilibrium values – meaning the explanation is logically valid and adequate only if none of the exogenous variables change during the time required for the endogenous variables to reach those values.

exogenous variable is to assess the hypothetical causal role of that one exogenous variable in the model. It can be considered to have a causal role only because it is the only exogenous variable being changed.

The use of the assumption of *ceteris paribus* is a common reason for criticizing Marshall's mode of explanation. Piero Sraffa [1926] famously criticized Marshall's reliance on partial equilibrium analysis and its related comparative statics analysis which deals with one variable at a time, rather than relying on general equilibrium analysis which deals simultaneously with all of the variables in the equilibrium model. Sraffa's criticism also was about the role of imperfect competition in the real world that we can see out our windows.

Over the last ninety years many critical arguments have been advanced that seem to suggest we spend too much time analysing equilibrium states. What these critics claim is that we should instead be worrying more about everyday disequilibrium phenomena before any equilibrium has been reached. Doubts about basing all economics on the concept of equilibrium can easily stem from the analysis of the necessary conditions for equilibrium attainment regardless of how the equilibrium is reached or analysed. Or as Arrow simply noted in the above quotation, it can stem from the question of who changes the price when the market is not in equilibrium.[15]

The microeconomics model builders' general concern for the possible limitations of equilibrium models will be the main topic in the remainder of Part I. The concern is mainly about what goes on in a state of disequilibrium, a concern which has two different sources. The first source of interest started with an article by Ragnar Frisch [1936] and another a little later by Paul Samuelson [1947/65, chapter IX], and (as I briefly discussed in the Prologue) was subsequently explained as a general problem facing all formal economic equilibrium models by Arrow in his 1959 article. Arrow just argued that equilibrium models must include some explanation for the necessary price adjustment that would lead to equilibrium attainment. As will be explained in Chapter 2, in effect Arrow identified a possible contradiction between the assumptions used to explain the behaviour of individuals in a state of equilibrium and the assumptions necessary to explain the adjustment of prices in a state of disequilibrium. While perfect competition may be consistent with any state of equilibrium, a disequilibrium state requires an explanation of the movement toward equilibrium. In this regard, Arrow suggested basing it on what textbooks see as the behaviour of imperfect competitors, which textbooks see as price setters. The second source of interest in disequilibrium economics can be traced back to the 1970s when there was a concern for the knowledge requirements of any participant in a state of equilibrium.[16] As I mentioned in the Prologue, the need to address the knowledge requirements for any state of economic equilibrium was not new since Hayek [1937] explicitly argued that

15. As I noted in the Prologue, early examples about these doubts include the three articles Arrow [1959], Clower [1959] and Richardson [1959] which will be the subject in turn of each of the next three chapters.

16. See Barro and Grossman [1971] as well as Solow [1979].

equilibrium model builders must clearly explain how the individual whose behaviour is explained in any equilibrium model acquires the knowledge needed
for any claimed state of equilibrium to exist. And I noted that later in 1959
from different perspectives, this need was also recognized independently
by Clower and by Richardson. As I said above, Chapter 3 will be devoted to
a discussion and explanation of Clower's perspective and Chapter 4 will be
devoted to discussing and explaining Richardson's 1959 critical perspective.
Richardson goes further to undermine all general equilibrium models of perfect competition. Before discussing all the problems with price adjustment
and equilibrium attainment, I will close this chapter by briefly discussing
Hayek's concern for the knowledge requirements for a state of equilibrium –
general or market.

## 1.3. EQUILIBRIUM AND NECESSARY KNOWLEDGE

In his 1937 article, 'Economics and knowledge', Hayek specifically argued that
an equilibrium model builder must provide some explanation for how any necessary knowledge is 'acquired' to reach a state of equilibrium.[17] In this regard,
he said [p. 45],

> The statement that, if people know everything, they are in equilibrium is true
> simply because that is how we define equilibrium. . . . It is clear that, if we want
> to make the assertion that, under certain conditions, people will approach that
> state [of equilibrium], we must explain by what process they will acquire the
> necessary knowledge. Of course, any assumption about the actual acquisition of
> knowledge in the course of this process will . . . be of a hypothetical character.

Hayek's 1937 view was not presented as any reason to drop using equilibrium
models in economic explanations but just as a suggestion that there is more
work to be done. Nevertheless, it is easy to see that the knowledge requirements to assure the attainment of a state of economic equilibrium are an easy
target for critics of neoclassical equilibrium model building.[18]

The logical requirements for the existence of a state of equilibrium have
specifically been the focus of many critics of neoclassical economics – much
less of the critics' focus has been on what is required for an individual's disequilibrium learning process. Some critics have argued that there is a problem

---

17. Note that throughout this book I am discussing explanatory knowledge, that
is, knowledge that yields explanations and understanding. Moreover, I am not discussing so-called 'know-how', or knowledge of languages which one might quantify by referring to the size of one's vocabulary. This issue will be further discussed in Chapter 6.
18. Parts of this section are revisions of some of what I argued in Boland [1986].

with any equilibrium concept which requires universal maximization. Many argue that it is unreasonable to think that any individuals going to the market could have acquired sufficient knowledge in advance to ensure they are all actually maximizing.[19] From a different perspective, others argue that even if such knowledge acquisition were logically possible, it would be difficult to justify given the limited cognitive abilities of human beings.[20] In either case the likelihood of ever being able to satisfy the knowledge requirement for equilibrium attainment is at least open to question.

As I will discuss in Chapter 4, the knowledge requirements for a state of equilibrium were a specific concern of Richardson, who in this regard identified two types of knowledge in his 1959 article: *primary* knowledge of one's own personal circumstances such as income, tastes, technical abilities, etc., and *secondary* knowledge such as what other people will publicly demand or supply in the market. While hardly anyone today would question assuming that everyone can know with adequate certainty his or her private circumstances, there is little reason to assume that everyone can have adequate knowledge about the public behaviour of the other market participants and particularly their future behaviour.[21] However, today few would doubt that every individual market participant can form expectations about the public circumstances. But Richardson says that unless there is some way of forming these expectations 'rationally' – that is, in a logically adequate manner – there is no reason to expect any individual to make the optimal choice in the market and hence no reason to expect an equilibrium ever to be reached.

All too often those interested in building an explanatory equilibrium model will simply begin with the assumption that the model's equilibrium exists. But unfortunately with such an assumption, these model builders are implicitly assuming that everyone has somehow acquired sufficient knowledge to be maximizing. Some do this without necessarily claiming that knowledge is perfect, but some critics would claim that unless the knowledge is perfect, a problem remains. Today this problem is often handled as a question of how to deal optimally with uncertainty – that is, with the question of whether there is an optimal method of making decisions when the truth status of one's knowledge is uncertain. How one answers this question depends on one's theory of knowledge or one's theory of learning. It is all too tempting to believe that everyone learns only by collecting more information – that is, by

19. This has been repeatedly argued by George Shackle; see, for instance, Shackle [1972].

20. This is the argument often attributed to Herbert Simon; see, for instance, Simon [1979].

21. Of course, one could build a model in which all individuals are identical or at least can be perfectly represented by a single agent and thereby try to obviate this problem – when it is just a matter of model building.

*induction* – in that more information means less uncertainty.[22] If one assumes inductive learning is possible – that is, that one learns just by accumulating many observations – then one could easily claim that the question obviously concerns the economics of information. It turns out that George Stigler in his 1961 article, 'The economics of information', argued just this. He claimed it is always a question of comparing the benefits and costs of information acquisition. Given that information collection is costly, Stigler claimed there to be an optimum degree of uncertainty such that the benefits of less uncertainty do not exceed the extra cost of reducing uncertainty. From his perspective of the economics of information, the quality of one's knowledge or expectations is chosen 'rationally' when the net benefit of information collection has been maximized. Things went further in the 1970s when rational expectations models were developed that simply presumed that learning is inductive, and consequently (as Stigler would have said), if a more perfect equilibrium were a benefit to anyone there would be an incentive for someone to collect the required additional information. As will be explained in Chapter 9, in rational expectations models perfectly certain knowledge is not logically required for a real-time equilibrium to exist despite what some of the critics of equilibrium models seem to think. Some critics of equilibrium model building will nevertheless claim that, in the absence of perfect knowledge, it is a disequilibrium that is being modelled. But given the presumed inductive learning, the builders of equilibrium models based on an assumption of rational expectations can easily explain away any so-called disequilibrium by one of two arguments. On the one hand, in the spirit of Stigler's approach, it can be argued that the charge of 'disequilibrium' wrongly presumes that the alternative perfect state of equilibrium is economically feasible. On the other hand, it can be argued that the state of disequilibrium exists only because not enough time has been allowed for the participants to acquire the necessary degree of certainty to make equilibrium decisions. As I will note in Chapter 6, if we were to follow Adam Smith's friend David Hume and accept his denial of the possibility of inductive learning and induction-based explanation, it is not clear that a state of disequilibrium can so easily be explained away. I will extensively return to these critical matters of knowledge and learning in Chapters 6, 12 and 15.

22. A brief discussion of this issue of methodology is necessary to note here. The idea of 'induction' is a much misused concept in economics. Despite what some may think, induction does not mean any explanation in which empirical or observational statements have been used. After all, explanations are exercises in deductive knowledge for which necessarily some but not all of the premises of the explanation are assumptions of a general nature (e.g., all consumers are utility maximizers). Induction as a means of explanation and learning is supposedly a process in which one makes many observations and tries to identify a pattern without assuming anything of a general nature. Such so-called inductive learning is easily dismissed as jumping to conclusions. Moreover, as I have explained in Boland [2003, ch. 1], there is no such thing as an inductive logic let alone an inductive proof.

## 1.4. MARSHALLIAN TEXTBOOK EXPLANATIONS
## VS. MODERN ECONOMIC MODEL BUILDING

I have discussed Marshall's view of equilibrium-based explanation only because that is what is utilized in typical Economics 101 textbooks and classes. Such a discussion is needed because this is what all economics students are expected to know and this includes today's equilibrium model builders since they virtually all began their education in economics learning Marshall's view of equilibrium explanation. In the rest of this book I will be critically examining what modern equilibrium model builders seem to think goes beyond Marshall's view. What goes beyond involves attempts to overcome many of the recognized shortcomings of Marshall's view of equilibrium model building.

# CHAPTER 2

Ↄᐯᕐ

# Equilibrium attainment
# vs. equilibrium necessities

Kenneth Arrow in his famous 1959 essay, 'Towards a theory of price adjustment', considered the main issue involved in any equilibrium model
built to explain prices – such a model must involve some notion of equilibrium
attainment. About this he argued [1959, p. 41]:

> [T]here exists a logical gap in the usual formulations of the theory of the per-
> fectly competitive economy, namely, that there is no place for a rational decision
> with respect to prices as there is with respect to quantities! A suggestion is made
> for filling this gap. The proposal implies that perfect competition can really pre-
> vail only at equilibrium. It is hoped that the line of development proposed will
> lead to a better understanding of the behavior of the economy in disequilibrium
> conditions.

This logical gap exists mostly for those who are limited to building equilibrium
models of perfect competition. What is the alternative? As we will see in this
chapter, Arrow provided an alternative in his article but as we will also see in
the next chapter, it has problems, too.

Before considering Arrow's 1959 article, let us step back and instead consider what I noted briefly in Section P.2 of the Prologue – namely, that neoclassical economics uses only one behavioural assumption, and when building equilibrium models an additional assumption of a state of equilibrium is
redundant. Of course, this redundancy is particularly true for those theorists
building general equilibrium models where the main mathematical exercise is to determine a set of equilibrium prices for all products and productive resources such as labour that would allow all market participants to be

maximizing. And if everyone is maximizing, then of necessity all markets are clearing by the usual textbooks' definition of the demand and supply curves – and thus there would be no need to add an assumption of the existence of a state of equilibrium. The problem here is that if the equilibrium model builder does not explicitly add the assumption of the existence of a state of equilibrium, the question can be raised concerning how the state of equilibrium comes about. And this state of affairs reflects the issue that in textbooks' general equilibrium models there is no built-in explanation for how prices adjust to the sought-after set of equilibrium prices. That is, for the textbooks' general equilibrium models there is no discussion of the process of price adjustment during any state of disequilibrium that would exist prior to reaching a state of equilibrium, beyond hand waving or inventing an imaginary auctioneer. Instead, in textbooks it is just assumed that all market participants are equilibrium 'price takers' – and this is predicated on the presumption of perfect competition whereby there is a sufficiently large number of market participants that every individual is a 'small fish in a big pond' and thus no individual demander or supplier can affect the price. Textbooks rarely recognize the need to address the question of how prices adjust to equilibrium prices, but instead rely on the fact that changes would cease when the equilibrium is eventually reached so that every price taker would be facing only equilibrium prices.

In his 1959 article, Arrow observed that our recognition that some buyer or seller would have to alter the 'given price' when it is not an equilibrium price means that we are not assuming that individuals are always price-takers as assumed in the perfect-competition theory of prices and quantities that is found today in most elementary textbooks. As I noted in the Prologue, Arrow suggested that what is required is some form of an imperfect-competition theory of price-quantity adjustment behaviour.

## 2.1. PRICE ADJUSTMENT IN A FORMAL MODEL

Price adjustment is still seen as a problem to be solved when building equilibrium models. As Franklin Fisher [1981, p. 279] puts it some time ago, 'One of the central unsolved problems of microeconomics is the construction of a satisfactory adjustment model which converges to equilibrium as a result of rational behavior on the part of the decision-making agents involved'. But he went on there to observe, 'This is not a simple matter, particularly when production and consumption are allowed to go on outside of equilibrium'.

With all of this discussion of the logical requirements for any equilibrium explanation, it is time to consider a typical model of a single market's equilibrium and see how all this plays out. Think of a single market of the usual textbook variety where the demand curve is downward sloping and the supply curve is upward sloping and where all participants are price takers. Let us follow the lead given by Arrow in his 1959 article and still found in many current textbooks. To do so we represent this market with two equations, one for the demand, $D$, and the other for the supply, $S$, as follows:

$$D = f(P, R), \qquad [2.1]$$

$$S = g(P, K) \qquad [2.2]$$

where $P$ is the going market price (which might not be the equilibrium price), $R$ somehow represents the exogenous income (or wealth) available, and similarly $K$ represents the exogenous allocation of physical capital to the producers. The two equations represent, respectively, the demand and supply quantities that would occur whenever everyone is maximizing utility or profit for the going price, $P$, and the exogenous givens, $R$ and $K$.

Those equilibrium model builders who only want to know the equilibrium price will simply equate $D$ and $S$ and then solve for $P$ given the values for $R$ and $K$ as well as the two functional relationships f(·) and g(·).[1] Formally, this presumed equality merely means a third equation is added representing a state of equilibrium:

$$D = S \qquad [2.3]$$

Not much can be learned from the solution for the price $P$ unless, as Arrow suggested, there are reasons given for why equation [2.3] should be true. We have reasons for why equations [2.1] and [2.2] are true – namely, all individuals are assumed to be maximizing and the two equations are merely logical consequences of such simultaneous maximization by all demanders in one case and all suppliers in the other.

Every elementary textbook illustrates these three equations simply by providing the usual market equilibrium diagram such as Figure 2.1 where, of course, point E is the equilibrium point represented by equation [2.3]. It is easy to prove that equation [2.2] can represent an upward sloping supply curve by simply noting that a profit maximizing firm produces where its marginal cost[2] is rising, which occurs where the production function used to produce the quantity supplied exhibits a diminishing marginal product of labour.[3] In perfect competition, to maximize profit the firm chooses the level of supply

---

1. Of course, these two functional relationships require explicit mathematical formula designed to represent implications of everyone maximizing.

2. For those readers who do not remember their Economics 101 class, I explained this jargon in note 4 in the Prologue.

3. For those unfamiliar with the elementary economics textbooks, the production function is a mathematical function representing the production of the commodity being supplied using an input of labour and what is called capital (representing

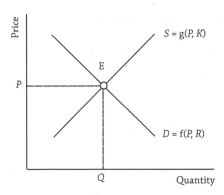

**Figure 2.1.** A market equilibrium

at which its marginal cost equals the given market-price. But, unlike proving that all supply curves are positively sloped, proving that all demand curves are downward sloping has always been problematic.[4]

In effect the problem that Arrow in his 1959 article raises concerns the completeness of our textbooks' equilibrium models of price determination. Specifically, if the model consists of only the three equations [2.1], [2.2] and [2.3], it still does not explain why equation [2.3] is true. Traditionally, textbooks rely on some unspecified price adjustment process to correct for any discrepancy in equation [2.3]. Arrow's term 'price adjustment' usually means equilibrium model builders must deal with how fast and in what direction the price changes. Speed of adjustment is usually represented mathematically by a calculus derivative and its sign (positive or negative) represents the direction of adjustment. As clock or calendar time ($t$) advances, the speed and direction of price adjustment depend on the amount of excess supply ($S - D$) and the adjustment can be represented by the following equation:

$$dP \, / \, dt = h(S - D) \qquad\qquad [2.4]$$

machinery). But, as I explained in Chapter 1, usually labour is the only variable input by virtue of the textbook's Marshallian definition of the short-run and usually the textbook diagram is just about a short-run market. And so, the usual reason for marginal cost to be rising is just the cost of a rising extra amount of labour needed to produce each extra unit of output. That is, as more labour is employed to increase supply, the diminishing amount of output produced by each extra amount of labour means that more extra labour is needed for each extra unit of output as the level of output and thus the labour employed and its total cost increases.

4. Long ago, it was recognized that there are many reasons for the difficulties. For one, it is claimed that some demand curves may exhibit so-called Giffen effects that can yield upward sloping demand curves; see Boland [1992, ch. 14]. For the purpose of Arrow's elementary discussion such complications are ignored.

What Arrow is suggesting is that any formal market equilibrium model must include an equation [2.4] or something equivalent. Equation [2.4] represents a situation where there can be a reason to adjust the price. In the case where demand exceeds supply, not all demanders are able to fulfil their chosen demand quantities (quantities that would maximize utility so as to be represented by equation [2.1]) simply because the store shelves are empty. In the case where supply exceeds demand, not all firms are able to sell their chosen supply quantities (quantities that would maximize profit so as to be represented by equation [2.2]). But, equation [2.4] by itself does not complete the formal equilibrium model of price determination. For it to complete the equilibrium model certain conditions about it must be assured. Specifically, to assure the equilibrium is attained, it is presumed that whenever equation [2.3] is true, $dP/dt$ equals zero; it is also presumed that when $S$ is greater than $D$ the price will fall. These presumptions can be formally represented as:

$$d(h(S-D))/d(S-D) < 0 \text{ and} \qquad [2.5]$$

$$h(0) = 0. \qquad [2.6]$$

For years, many equilibrium model builders might have been satisfied to simply presume that [2.4], [2.5] and [2.6] are all true, and thereby presume to have completed the reasoning[5] for why equation [2.3] is true. But, where does the formula $h(\cdot)$ come from? As already noted, the formulas $f(\cdot)$ and $g(\cdot)$ explicitly represent explanations using the same behavioural assumption, namely, maximizing – the singular neoclassical behavioural assumption. And since there is only a single behavioural assumption, $h(\cdot)$ needs to similarly represent an explanation using some sort of a behavioural maximizing assumption.[6]

## 2.2. EQUILIBRIUM ATTAINMENT
## AS AN EXPLICIT PROCESS

It is important to recognize that the question of price adjustment or equilibrium attainment needs to be considered an out-of-equilibrium process.

5. As a matter of jargon, completing such an explanation is sometimes called 'closing the model'.

6. It is not difficult to see that there is nothing here that tells us how long it would take for the price, $P$, to equal the price for which equation [2.3] is true (for the given equations [2.1] and [2.2]). Whenever demand exceeds supply, if the condition [2.4] is specified such that the price never rises fast enough to cause the excess demand to become excess supply before the equilibrium is reached, $(S-D)$ and $dP/dt$ might both approach zero but only as $t$ approaches infinity. In other words, while the price can change in the right direction, it may easily be that the equilibrium is never reached in real time.

As Yves Balasko puts it, 'It is implicit in the definition of competitive equilibrium that competition exerts some forces on out-of-equilibrium prices that prevent them from remaining stationary' [2007, p. 413]. Moreover, as Arrow observed, 'The failure to give an individualistic explanation of price formation has proved to be surprisingly hard to cure' [1994, p. 4].

In his 1959 article Arrow did actually addresses this fundamental problem but, judging by most textbooks, what he suggests has not caused any major revolutions in the practice of equilibrium model-building methodology in economics. To appreciate his suggestion for addressing what it takes to complete a formal equilibrium model of price determination, consider a textbook market of $m$ buyers that are represented collectively by [2.1] and $n$ sellers represented collectively by [2.2]. At any given price each participant decides either how much to buy to maximize utility or how much to sell to maximize profits. The total demand at any given price is merely the sum of all that is demanded by the $m$ consumers and similarly the total supply is merely the sum of all that is supplied by the $n$ firms at that given price. If the given price is actually the equilibrium price, the total demand will exactly equal the total supply. In such an equilibrium state each individual needs only to consider the given price and his or her private circumstances (income, resources, technology, etc.). Should the price they are given be an equilibrium price they will all *unintentionally* choose the quantities which will be market clearing – regardless of the number of buyers and sellers. But, what happens if the market participants are not given an equilibrium price? This is the question that equation [2.4] purports to answer in that it considers the existence of an excess demand or excess supply resulting from the going market price not equalling the price that would be obtained when equation [2.3] is true. Whenever [2.3] is false, the market is said to be in a state of disequilibrium. To see this, let us think of a market place where sellers enter from the door on the right with their quantity to sell and demanders enter from the door on the left and in each case one at a time as needed to complete a pending transaction.[7] They all are facing the same given price since the textbooks' market usually presumes everyone is a price taker. In this market, a transaction only involves coming to agreement as to the quantity to purchase or sell. As each demander enters he or she determines if the last seller who entered still has enough unsold to fulfil his or her desire to maximize utility. If not, the demander waits for the next supplier to enter. Similarly, as each supplier (or its agent) enters, it determines if the last demander still has some unfulfilled need. To visualize the situation, consider Table 2.1 for any given market with everyone facing the same going price ($P$):

7. For reasons that will be apparent below, we have to assume no one leaves the market place until everyone has entered and the equilibrium has been reached.

**Table 2.1.** *THE MARKET SITUATION*

| | Demand | Supply |
|---|---|---|
| | $d_1$ | $s_1$ |
| | $d_2$ | $s_2$ |
| | $d_3$ | $s_3$ |
| | . | . |
| | . | . |
| | . | . |
| | $d_m$ | $s_n$ |
| Totals: | $(d_1+d_2+\cdots+d_m)$ | $(s_1+s_2+\cdots+s_n)$ |

*Note:* The subscripts identify individual demanders or suppliers.

Now consider such a sales situation where the last supplier who wishes to sell the product in the market place may find nobody coming in the other door even though this supplier has not sold enough to be able to maximize its profit. For this market's situation, it is usually assumed that no actual transactions (or trades) take place until there is nobody being unable to maximize.[8] Of course, whenever the given price (P) is not the equilibrium price – that is, at the going price P, the 'total demand' does not equal the 'total supply' – no transactions will take place at P. For the posited sales situation being considered, this is because total supply exceeds the total demand. And, of course, such an excess supply merely means that at least one of the n suppliers cannot maximize profit. In such a case, equilibrium models must recognize that someone (the disappointed supplier here) will have to compete by offering a different selling price if maximization is still the objective. Since I am talking about a market where there is excess supply, any disappointed seller would have to bid the price down to induce some buyer to purchase from this supplier instead of some other supplier.[9] Obviously this now leaves that other supplier in the same position of not being able to sell its profit maximizing output – as a result this supplier must also bid the price down. This back-and-forth continues in such a competitive bidding manner until the market's excess supply is reduced to zero. And thus the bidding process stops at a lower, equilibrium price and the pending transactions are allowed to take place.[10]

8. For my example, think of no transactions taking place but the previous demanders and sellers who have entered so far are just waiting around to complete a transaction. Only when everyone is able to maximize can the waiting transactions take place.

9. And note that, as I said in the Prologue, the reason is the same behavioural assumption – this time used in reverse. That profit is not being maximized is the incentive to alter the going price.

10. And similarly where there is excess demand, any disappointed demander would have to bid the price up so as to induce one or more of the sellers to sell to him or her

## 2.3. EQUILIBRIUM VS. IMPERFECT COMPETITION

Arrow observes that the recognition that either a buyer or a seller would have to alter the 'given price' when it is not an equilibrium price obviously means that individuals are not presumed to be price-takers, as would be the case in the text-books' perfect-competition theory of prices and quantities. What is required, Arrow said, is some form of an imperfect-competition theory of price-quantity behaviour. The question being posed here is, how can the adjustment of price be explained with the singular behavioural assumption of neoclassical economics?

In his 1959 article Arrow does consider an obvious alternative view of price determination [pp. 43–4]:

> Before discussing the mechanics of price adjustment under competitive condi-
> tions, we may consider the determination of price under monopoly. Here, there
> is no question of the locus of price decisions. In the standard theory ... the
> monopolist fixes his price and output to maximize $R(x)-C(x)$, where x is output,
> $R(x)$ the total revenue curve, and $C(x)$ the total cost curve. Price and output are
> related by the demand curve, and the firm's output will, therefore, always equal
> demand. This theory clearly presupposes that the monopolist knows the true
> demand curve confronting him.

Depending on whether the market is exhibiting excess demand or excess supply, Arrow suggested considering the last demander to be a monopsonist[11] or the last supplier to be a monopolist. If model builders do this, then they can use the imperfect-competition model to explain price determination.[12] In effect and most important, this means Arrow is saying there is one theory of price determination when demand equals supply and a different theory of price determination when demand does not equal supply.[13]

In simple terms such as those discussed with Table 2.1, this means in the case of excess supply for any point on the demand curve, the last seller will

instead some other demander. And this means that this other demander will have to respond by bidding the price further up and thus the price goes up in such a competitive bidding manner. Bidding the price up would continue until the excess demand is thereby ultimately reduced to zero such that the price stops rising at a higher, equilibrium price.

11. This jargon refers to a market situation in which there is only one demander – see further Arrow [1959, p. 43].

12. There is, however, a major problem with this strategy for explaining the price determination, since unless this last seller's supply is extremely small, the producer will be operating at a less than efficient output level – as will be explained in the next chapter, one at which average cost is not lowest (unlike in the case of a long-run equilibrium in a perfect-competition model).

13. It should be noted that textbooks could easily recognize that suppliers can keep a standing inventory such that the bidding urgency would be much less. But doing so only begs an additional question of what is the optimum inventory size.

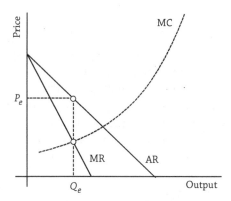

**Figure 2.2.** Profit maximizing imperfect competitor

choose to produce a supply quantity such that its marginal cost (MC) equals marginal revenue (MR) as illustrated in Figure 2.2 and since it is not perfect competition, the marginal revenue will be below the market price simply be-cause the usual textbooks' demand curve is negatively sloped.[14] And for a given demand curve, Arrow presumes that the supplier will eventually find the equilibrium quantity to supply, either directly or perhaps by trial and error. After providing another set of equations for formally representing the 'rules' (i.e., necessary conditions) for the monopolist to be maximizing profit, Arrow does add a significant observation: 'These rules have concealed in them im-plicit assumptions about the monopolist's knowledge of the demand curve facing him (I assume he has complete knowledge of his cost curve)' [p. 44]. He then – referring to Oscar Lange's work on developing a theory of price ad-justment for monopolists – also recognized that the 'monopolist must know the demand at the price chosen'. Moreover, to make the maximization rules operationally meaningful (i.e., falsifiable), the monopolist 'must know the elas-ticity of demand at that price'.[15] However, 'the monopolist presumably does

14. Actually, whenever the average revenue curve (which is just the demand curve itself) is falling, marginal revenue (MR) will necessarily be less than average revenue (AR) as shown in the diagram. Note that I am saying less than rather than falling as the MR does not have to be falling when the AR is falling – it just needs to be lower. Textbooks that always show demand curves as straight lines mislead students into falsely thinking the MR must also be falling in this case. The common textbooks' use of a straight-line demand curve will be an important issue in the next chapter.

15. For those unfamiliar with this economics jargon, 'elasticity of demand' is a measure of the responsiveness of demand to a change in the going price. For example, whenever demand elasticity is unitary, a 10% increase in the price will be followed by a 10% reduction in the quantity demanded which means the amount of the budget spent on good $X$ does not change and hence there is no necessary change in the budget spent on good $Y$.

not know the entire demand curve, for otherwise he would jump immediately to the optimal position' [p. 44].

I raise all this to indicate that the whole notion of a market consisting only of price takers is somewhat suspicious at best when it comes to understanding the *process* of reaching the market's equilibrium. The problem, though, is that the mathematical characterization does not correspond to realistic situations it is supposed to represent. Food for thought hopefully.

But, as we will see in the next chapter, it is not clear that Arrow's suggestion for dealing with the disequilibrium situation by invoking what is taught in those textbooks' chapter on the behaviour of a monopolist or imperfect competitor is a as promising as Arrow seemed to think.

# CHAPTER 3

⌒᷍⌒

# Does general equilibrium attainment imply universal maximization?

This chapter will examine the suggestion Arrow provided in his 1959 article for assuring an explanation of equilibrium attainment – namely, that some form of imperfect competition must be considered.[1] About Arrow's suggestion, Jean-Pascal Benassy has noted [1976, p. 69]:

> Monopolistic price setting and equilibrium have had quite a long history in Economics . . ., but still most contributions to general equilibrium theory continue to view the firm as a price taker. Recently Arrow [1959], noticing that 'there is no one left over whose job is to make a decision on price' advocated a more realistic approach to price determination with firms behaving monopolistically, and stressed particularly the relation between monopolistic and out-of-equilibrium behaviour.

As will be seen in this chapter, unlike the usual model of perfect competition in which market participants need only know the going price to make their maximizing decision, if we follow Arrow's suggestion and consider, instead,

---

1. For readers unfamiliar with standard textbook jargon, beyond what I said in note 1 of the Prologue, imperfect competition means: there is more than one producer in a market in which all are producing the same commodity but there are not enough producers to make everyone a price taker. A related type of imperfect competition occurs where the firms are producing competing brands for similar but not identical commodities – think of the automobile industry. The name for this type of imperfect competition is monopolistic competition since each firm has a monopoly over its brand. Here we will be discussing only imperfect competition and its extreme form when there is only one producer – which is called a monopoly.

imperfect competition, the going price may not be enough information. As Clower observed [1959, p. 705]:

> Although no one doubts the importance of limited information and uncertainty as factors affecting the management decisions of actual business firms, economists have long been reluctant to introduce such complications into traditional models of price and output determination . . . the majority of professional economists, . . . continue . . . to think and work in terms of models which presuppose a world of perfect information, perfect certainty and instantaneous response to changing circumstances.

As we saw in the previous chapter, the primary point addressed in Arrow's 1959 article was the completion of any formal equilibrium model that purports to explain the determination of the market price as an equilibrium price. The usual singular behavioural assumption is sufficient to allow us to explain demand curves and supply curves and, if we include Arrow's equation for explaining the disequilibrium price adjustment, we can at least explain the direction of any disequilibrium price adjustment as well as the speed of the adjustment. But it must be recognized that without a model that indicates how that adjustment is determined with our singular behavioural assumption, we may have a mathematically adequate model but one that is incomplete as an explanation of equilibrium *attainment*. Arrow suggests that our textbooks' model of the imperfect competitor does show that an equilibrium price can be reached, but it is not an explanation for how long this would take – even in the trial-and-error manner he suggested. So the price adjustment theory that Arrow seeks is incomplete even though we know where the price will end up eventually.

### 3.1.  THE EQUILIBRIUM ACTUALLY REACHED
### BY AN IGNORANT MONOPOLIST

In the case of an imperfect competitor, it turns out that even if we did have such an explanation of the speed of adjustment and how long it would take to attain an equilibrium price, there are reasons to suspect that a complete model might not result in the usual profit maximizing equilibrium price as some might think. Although Clower's 1959 article does not address the question of the speed of adjustment, it does address the question of whether a complete model of the imperfect competitor does yield the textbooks' profit maximizing equilibrium price. In so doing, he happens in effect to identify another problem for Arrow's desired complete equilibrium model of price adjustment. He identifies this problem in three steps of model building. In each of the three steps, he addresses a model of a monopolist rather than merely an imperfect competitor.

The first step is his Model I where the monopolist has full prior knowledge of the demand curve it faces[2] and so knows the market's true marginal revenue (MR) curve. Knowing the MR curve, the monopolist looks for the supply quantity at which the MR curve crosses the perfectly known marginal cost (MC) curve since at that quantity ($Q_e$) profit is being maximized. Having determined this level of output, it sets the corresponding price ($P_e$) indicated on the known demand curve (AR).[3] So, when the monopolist sends the $Q_e$ amount of supply to the market, the market clears at the resulting price $P_e$ as anticipated. All this is illustrated in Figure 3.1 (which is the same as Figure 2.2 from the previous chapter).

The second step is his slightly more dynamic Model II. Model II differs from Model I only by letting the monopolist be somewhat ignorant. Model II's monopolist does not have prior knowledge of the demand curve but, like Arrow's suggested imperfect competitor, is able to determine the profit maximizing price in a trial-and-error manner. Moreover, for this second model Clower also assumes that the demand curve facing the monopolist is the textbooks' downward sloping straight line. While the monopolist of Model II differs from the monopolist of Model I by being somewhat ignorant of the market's demand curve, both can still be represented with Figure 3.1. The key point about Clower's Model II monopolist is about how it differs from what he later presents in his Model III. The difference is that in his Model II the monopolist

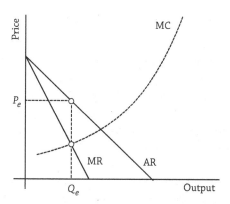

**Figure 3.1.** The knowledgeable monopolist

2. And since a monopolist is by definition the market's only supplier, the market's whole demand curve is what the monopolist faces.

3. For those unfamiliar with this jargon, the price (P) is always equal to the average revenue (AR) as a matter of algebraic definition. That is, total revenue (TR) is nothing more than the algebraic product of the price times the quantity sold (i.e., $P{\cdot}Q$) and average revenue is obtained by dividing the total revenue by the quantity sold ($AR = P{\cdot}Q/Q = P$).

learns the true demand curve's slope and position with this trial-and-error procedure but this not necessarily the case for the monopolist in his Model III.

Interestingly, Clower published a little known 1955 article that looked at the process of price determination. Instead of presenting the price-setting monopolist depicted in his 1959 Models I and II, Clower presented a purely competitive market with a 'market authority' which today might be a computer program. He called this person or machine a 'marketee'. This marketee would perform the process of price determination in a step-by-step manner involving estimating demand curves. It was a trial-and-error process whereby adjustments would be made should the estimated demand curve prove false. He then presented a monopolist which does basically what the marketee did to determine a market clearing price facing a demand curve that resembles what we see in textbooks – specifically, a straight line that is negatively sloped. With this he then derived an implicit supply curve for this monopolist that is upward sloping and located to the left of the monopolist's marginal cost curve. And the ultimate result of the trial-and-error process – a market clearing price – is reached where this implicit supply curve crosses the true demand curve.

Clower's 1959 Model II monopolist seems to behave as his 1955 monopolist did to find a market clearing price. The main point of these models was that the market-clearing price can be determined even when the marketee or the monopolist is ignorant of the true nature of the market's demand curve. In the 1955 article [p. 226], he concludes:

> The implications of the above argument should be clear: *the mechanical aspects of price determination processes are always the same regardless of market structure considerations.* Thus, any single-product oligopoly situation (monopoly, bilateral monopoly, duopoly, price leadership, etc.) can be described by a statical model which is formally analogous to a model of an isolated competitive market for a single commodity . . . What have been thought to represent separate branches of economic analysis are thus seen to be no more than parts of a more general, unified theory of price determination.

Interestingly, in the 1955 article, Clower seemingly ignores the fact that at the resulting market clearing price for the ultimate adjusted estimated demand curve, the monopolist is not actually maximizing profit if the estimated demand curve happens *not* to coincide with the true demand curve. So in the 1959 article Clower corrects this lacuna in his third step, which goes well beyond what he presented in his 1955 article. He does this by presenting an explicit specification of a model of a truly *ignorant* monopolist (his Model III) whose estimated demand curve does not at the end coincide with the true demand curve. That is, this ignorant monopolist differs from that in the first two models since this monopolist has no knowledge about the market's demand curve and in particular knows nothing about the shape, slope or location of that demand curve.

To better understand Clower's Model III of the ignorant monopolist,[4] one has to adopt Friedrich Hayek's or John Hicks' view of the process of decision making. Using their terms, Clower's monopolist should be thought to form a plan as a first step. That is, *before* going to the market with its chosen quantity of supply, the firm decides what the market price is expected to be. Perhaps the monopolist does this after a couple of prior tests of the market. Specifically, Clower [1959, p. 708] says

> If the demand function is assumed to shift not too frequently (no more often, say, than every three market periods), it is then permissible to argue that the monopolist will acquire *correct information* about the *slope* of the function from *ex post* price and sales data (the three-period scheme allows time for two consecutive observations so that a conjecture concerning slope can be made, and then time for a final observation to 'confirm' the conjecture!). This having been accomplished, the monopolist's only remaining problem is to find the *position* of the demand function in different periods of time.

Here he is talking about the monopolist of his Model II. In order to distinguish his truly ignorant monopolist from the one in Model II that eventually learns the true demand curve, the monopolist of his Model III must make assumptions about the shape and position of the true demand curve – but these assumptions can be false. To see how this matters, it is most important to not see the monopolist as deciding how much to produce as well as putting a price on the product. Instead see the monopolist only deciding how much to produce. The monopolist then just has an agent who independently transports the chosen supply quantity to the market and then lets the market determine the price – usually the one that clears the market that day. This way this monopolist has no direct way to actually learn the market's true demand curve and instead has to always make its supply decision based only on the reported previous day's clearing price. Based on that previous price, the firm decides how much to produce next – the 'how much' as usual is assumed to be the amount which would maximize profit as depicted in Figure 3.1 but where $P_e$ now represents the expected price. The next step is to go to the market with the consequences of that decision, that is, with the planned quantity for the expected price.[5] The last step depends on the outcome of the agent's trip to the

4. This requires going beyond what Clower explicitly presents in his 1959 article because, unlike his earlier version, Clower [1955], in 1959 he was unfortunately more engaged in demonstrating formal mathematical modeling techniques than in being more straightforward in his exposition of the ideas being presented.

5. The supply quantity is chosen so the MC equals the expected MR and the firm thus expects the price to be at the corresponding level ($P_e$) on the expected AR curve

market.[6] If this monopolist's expectations as to a market clearing supply were correct, then all that was produced will have been sold and the returned price will be exactly as expected.[7] If the market-clearing price were underestimated, the firm would realize that it did not choose a profit-maximizing supply quantity (and, as Clower [1955] puts this, the firm would have to reduce its inventory). If instead the firm overestimated the market clearing price, then it will be left with a quantity of unsold goods.

The most important idea introduced in Clower's Model III is that the monopolist must decide what to do in response to a failure to maximize solely because its expected price turns out to have been incorrect. The question now is: how does the monopolist learn from its refuted expectations? Since the monopolist of Model III does not directly know much about the market's demand curve beyond the prior refuted expectations, to interpret failed expectations requires making assumptions based only on a few limited bits of information, that is, on those reports of failed trips to the market. So, before forming an expectation about the next market's price, the monopolist must try to understand why its last expected price was wrong.

Clower refers to his monopolist as 'an ignorant monopolist' simply because this monopolist does not know the market's true demand curve's shape or location. And thus to make up for the ignorance of the market's true demand curve, Clower's monopolist must make a priori assumptions about the shape and location of the market's true demand curve. To make his point, however, Clower has his monopolist assuming that the demand curve faced is the one usually found in textbooks. To illustrate this, he characterizes his monopolist as assuming its shape is linear (i.e., a straight line) when in fact the true demand curve is not. As a consequence of this false assumption concerning shape, the monopolist mistakenly interprets each subsequent failed expectation as evidence of a change in location – specifically, a *parallel shift* in the linear demand curve (since this is the simple way to maintain the monopolist's linearity assumption and the deviations from the expected results). The presumed shifted demand curve is used to calculate a presumed shifted marginal revenue curve which is then used to identify the next profit maximizing supply quantity to send to the market. As usual, that quantity will be the one where the calculated marginal revenue curve intersects the known marginal cost curve.

in Figure 3.1. Remember that the AR curve is just the monopolist's expected market demand curve and hence the expected MR curve is the one corresponding to that expected AR curve.

6. Again, think of Clower's monopolist as having an independent agent who delivers the firm's chosen supply quantity to the market and then returns and reports to the monopolist what the resulting price was that cleared the market – that is, whether or not the expected price was confirmed.

7. Of course, selling all that was supplied could occur if the expected equilibrium supply was underestimated but this would just lead to a higher market price.

It would be easy to stop here and just say that such a process eventually ends up with the monopolist's decisions converging on a market clearing price – that is, on a market equilibrium – however doing so would miss the most important point of Model III. Specifically, by assuming a stable configuration of cost and demand curves, the firm can easily reach an 'equilibrium' at which the calculated marginal revenue curve is not the true marginal revenue curve and hence the firm is not truly maximizing profit when it chooses a supply quantity where this calculated marginal revenue intersects the true marginal cost as depicted in Figure 3.2.

This figure illustrates, I think, everything Clower's final model was intended to show.[8] In this market, the true demand curve is not the straight line as usually found in textbooks – even though Clower's monopolist assumes that it is. Clower's monopolist, having assumed that the market's demand curve is the textbooks' straight line, eventually reaches (by trial and error) a level of supply that clears the market at an equilibrium price. It does this by going back-and-forth day-after-day (or week-after-week) to the market making subsequent adjustments to compensate for failed expectations. The adjustments are those where his monopolist assumes the failure was due to unknown shifts in the assumed market's straight line demand curve rather than actual movements along the true and unmoving non-linear demand curve. So long as the true demand curve does not actually move, the back-and-forth activity eventually

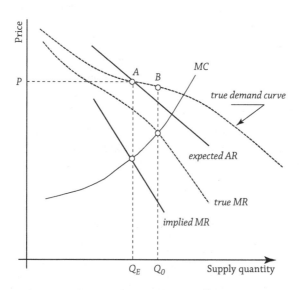

**Figure 3.2.** 'Equilibrium' for Clower's 'Ignorant Monopolist'

8. Actually, I showed this figure to Clower several years ago and he agreed that it does.

settles down to what is illustrated in Figure 3.2.[9] Specifically, the firm ends up supplying $Q_E$ and the market price $P$ results. This price ends up corresponding to point $A$ on both demand curves – the (dotted) true curve and the resulting assumed and expected demand curve (the solid line through point $A$ which is used to calculate the expected average revenue, $AR$[10]). The calculated implicit marginal revenue curve (the lower straight line which is implied by the firm's assumed demand curve) is the end result – and since this calculated implicit marginal revenue curve intersects the true marginal cost curve at $Q_E$, the firm thus thinks it is maximizing profit when, in actuality, it is not maximizing because the true marginal revenue does not intersect the true marginal cost at $Q_E$ but at $Q_0$!

The most important result, but one Clower does not make much of, is that contrary to the usual notion of a market equilibrium, the ignorant monopolist's market is in equilibrium (demand does equal supply) even though the monopolist is not really maximizing profit but, given its false assumption, thinks it is. This, of course, is because the monopolist is simply ignorant of the true demand curve. This is an important result since in textbook economics, universal maximization of participants in any market is always coincident with the state of market equilibrium. In Clower's Model III there is no reason for them to be coincident except by accident.

The reason for my discussing Clower's article here is that in his equilibrium Model III of an ignorant monopolist, the monopolist tries to learn from the available *disequilibrium information* provided by the market. But to do so, the information needs to be interpreted and such interpretation depends on the assumptions made by the decision maker. Moreover, what is also recognized is that Clower's decision maker must have some way of dealing with disappointed expectations concerning price or demand curve's slope and thus must be equipped to deal with such errors. In other words, information from a disequilibrium market needs to be interpreted and such an interpretation depends on fallible assumptions. If the assumptions are false, then contrary to what our usual textbook behavioural assumption would have us believe, there is no reason to think that since demand equals the chosen supply, the firm is truly maximizing.

9. And, of course, like Arrow's model, there is no explanation of how long it takes to reach the equilibrium price.

10. Keep in mind the elementary point that the market's demand curve at each quantity indicates the monopolist's average revenue (AR) since the price is always the calculated AR. Also keep in mind that the price will necessarily equal marginal revenue only by definition of a perfectly competitive market in which all firms are assumed to be price takers. For the price-taking producer it means that no matter the quantity produced, it will receive the value of the given price for each additional unit of output and thus marginal revenue is the given price and thus constant. This is why textbooks always say for the profit maximizing perfectly competitive firm, marginal revenue equals price.

Clower's article also puts into question Arrow's idea that by considering the textbook notion of an imperfect competitor, one has a ready-made mechanism to provide the needed means to complete Arrow's formal model of price adjustment. Recall, Arrow suggested that the textbook model of an imperfect competitor could tell us how price is determined even in a disequilibrium situation. But, from Clower's Model III (see Figure 3.2 above), one can see that things are not so easy since there is no explanation for why the imperfect competitor (let alone a monopolist) would know the demand curve it faces and thus no reason for it to know the true marginal revenue curve it faces and uses to compare with the marginal cost curve to find the level of output where profit is maximum.[11] Instead, as Clower demonstrates, the imperfectly competitive firm or monopolist must make assumptions about the marginal revenue it faces and any false assumption may lead to an equilibrium situation but one for which profit is not actually maximized even though the imperfect competitor or monopolist thinks it is. In short, disequilibrium information is not enough to assure maximization although it might be enough to assure equilibrium attainment.

## 3.2. AN EQUILIBRIUM STATE AS A SUB-OPTIMUM

A supplier's ignorance of the market's demand curve is not usually dealt with in the textbooks' theory of the firm because it is just assumed that for a perfectly competitive firm, there is no need to have knowledge of that demand curve. Perfectly competitive firms are simply price takers and that is all they need to know. The firm needs only worry about maximizing its profit given the going price. But as Clower observed in his 1959 article [p. 716]:

> Ignorance about demand conditions is a ubiquitous feature of market life in the real world. Whatever its source ... it is a factor to be reckoned with in any attempt to describe observed output and price behaviour. Moreover, the phenomenon of demand ignorance implies – in a plausible if not in a demonstrative sense – the existence of learning and response mechanisms through which output and price decisions arrived at on the basis of provisional conjectures are revised in the light of realised results and ultimately reconciled with experience.

11. It might be argued that the monopolist or imperfect competitor could survey or sample some demanders, but in Model III this would have to be done after each failed expectation given the presumed demand curve's shift. Moreover, of course, interpreting surveys and samples involve even more assumptions and possible errors. In any case, such an approach is not typical of the textbooks' equilibrium models of the firm.

The obvious alternative to assuming a state of equilibrium is to begin by assuming the economy is in a state of disequilibrium, which may or may not be temporary. If we do consider a state of disequilibrium we must be looking at a state where at least one individual is not maximizing and thus at a state which is sub-optimal. This observation gives new meaning to what Arrow was saying in his 1959 article. If the explanation of how prices adjust requires an analysis equivalent to imperfect competition, then just what is an equilibrium in an imperfectly competitive market? As Arrow recognized, the issue of imperfect competition does not usually involve the choices made by demanders. Instead, an imperfect competitor is characterized as a non-price-taking producer whose behaviour can have an affect on the price given for the market demand curve it faces. As such, the imperfectly competitive producer does not face a marginal revenue that is always equal to the market-given price as a textbooks' perfect competitor would face.

Since the publication of a 1933 book by Robinson, all neoclassical textbooks have been easily adding a chapter to explain such an imperfectly competitive firm, but with the same neoclassical behavioural assumption used with the perfect competitor. Such chapters show a market with few producers which compete for market share. And given there are a few competitors, these chapters clearly show an imperfect competitor's equilibrium as an output level at which marginal revenue equals marginal cost (because profit maximization is assumed) and at which total revenue equals total cost (because competition is assumed to be sufficient to eliminate excess profits[12]). Thus, in her presentation, which is used in most textbooks, the imperfect competitor can still be maximizing profit whenever the market's competitive equilibrium is presumed to be assured.

If we look at the typical textbooks' view of the firm in such an imperfectly competitive equilibrium, we will see the usual 'U-shaped' average cost curve with the marginal cost curve rising and intersecting at the lowest point on the average cost curve, as in Figure 3.3. We will also see that the firm's effective average revenue curve is downward sloping like a demand curve (since by the textbooks' definition, imperfect competition means each firm's output level affects the going price).

Being an imperfectly competitive market rather than a monopoly, the level of competition in such a market is always assumed to be sufficient to assure that excess profits are driven down to zero once the market's long-run equilibrium

---

12. As noted above in this chapter's first footnote, textbooks distinguish monopolies from imperfect competitors that compete for market share. For the latter there are enough competitors who through price competition or entry and exit in the market eventually can drive excess profits or excess loses down to zero. One cannot assume that such is the situation with monopolists, which is probably why elementary textbooks usually do not look kindly on monopolists.

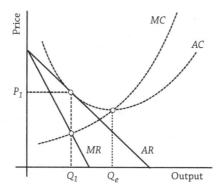

**Figure 3.3.** Imperfectly competitive long-run equilibrium

is reached. The zero-excess-profit equilibrium for a single firm implies its average cost (AC) curve is not only equal to its average revenue but it is also tangent to its average revenue (AR) curve at the level of output at which the marginal cost (MC) equals the marginal revenue (MR) and thus at the profit maximizing point. Since the average revenue curve is falling at that point, the marginal revenue is less than the average revenue, and thus the marginal cost is less than the average cost. And given this, the profit maximizing point is to the left of the lowest average cost – and that is the point where average cost is falling as shown in Figure 3.3. If all firms in the imperfectly competitive market are maximizing and selling the level of output that matches the level of demand, is this state of market equilibrium optimal from the social perspective of the whole economy?

While some may think that any complaint or criticism of explanations of prices based on imperfect competition can be overcome with some sophisticated mathematical models, I am not so sure. At a minimum, as Clower's 1959 article demonstrates, there may be a serious problem with any explanation of a state of equilibrium that relies on a supply decision maker who needs to know the demand curve in order to reach an equilibrium by attempting to maximize profit. And surely, if the Marshallian long-run equilibrium or equivalently a Walrasian general equilibrium involves some imperfect competitors, then some firms are not producing at an efficient level of output since, as shown in Figure 3.3, they are not producing where average cost equals marginal cost $(Q_e)$. That is, they are operating at $(Q_1)$ where average cost could be reduced by increasing output – this situation implies the firm is facing increasing returns were it to increase the scale of output.[13] This means that output is capable of being produced at an average cost (and hence average resource use) that could be lower. From the whole economy's perspective, more output could be produced with the same amount of the economy's available resources. From

13. Increasing returns (to scale) will be discussed more in Chapter 5 as well as in the chapters in Part II.

this perspective, one can see why some economists might see perfect competition – in which price equals marginal revenue – to be socially optimal.[14]

Clower's Model III of an equilibrium in which the monopolist is not producing at its optimal level – even though it thinks it is – illustrates an equilibrium that is not optimal for the monopolist as well as the whole economy. This is not like a model of an economy which includes imperfect competitors that are actually maximizing (Figure 3.3) – for which, it has been argued, a Walrasian general equilibrium is still possible.[15] Such a general equilibrium allows everyone to be deliberately maximizing but such an equilibrium would still be socially sub-optimal for the reasons provided in the previous paragraph. But Clower's Model III of an equilibrium state does not entail everyone truly maximizing as would necessarily be the case for every socially optimal Walrasian general equilibrium model.

14. It could instead be argued that if one considers the large amount of costly time required to learn about all of the huge number of competitors needed for the market to exhibit perfect competition (large enough to assure everyone is a price-taker), having consumers face an imperfectly competitive market might be the socially optimal situation. That is, if one adds the consumer's cost of search to the price of a commodity, the consumer's cost implications for the market reaching an imperfectly competitive price may be lower than for the perfectly competitive price given the lower number of producers in the imperfectly competitive market.

15. See Hart [1985] and Gabszewicz [1985].

# CHAPTER 4

༶

# Time and knowledge matters
# for general equilibrium attainment

A theme throughout this book will involve the knowledge and information needed by any equilibrium model's participating individuals and
this theme is central to this chapter's consideration of Richardson's 1959 article. The matter of information was also of interest to Stigler who observed,
'One should hardly have to tell academicians that information is a valuable
resource: knowledge *is* power. And yet it occupies a slum dwelling in the town
of economics' [1961, p. 213].

The main issue so far has been the requirements for completing a formal
market equilibrium model. Chapter 2 was about Arrow's 1959 article and his
argument that such a model must include an assumption that represents
some sort of dynamics of price adjustment, and that it is not enough just to
assume demand equals supply in all markets. Chapter 3 was about Clower's
1959 article and its discussion of how an ignorant monopolist would decide
on what price to expect prior to reaching a market equilibrium. In both cases,
the discussion was about attaining a state of equilibrium in a market model.
Moreover, it is important to recognize that both Arrow and Clower were discussing equilibrium states that are attained as the result of actions of or choices
made by autonomous, independent individual consumers or producers.

In this chapter I will be considering Richardson's 1959 article and its model
of general equilibrium attainment, which, like Clower's examination of a
model of a monopoly in equilibrium, also sees the need to address the knowledge requirements of an equilibrium model if the equilibrium is presumed to
be attained in due time.

## 4.1. KNOWLEDGE AND LEARNING
## IN ECONOMIC MODELS

Historically, the key consideration of a role for learning and knowledge acquisition was the recognition that, to quote Marshall, 'perfect competition requires a perfect knowledge of the state of the market' [1920/64, p. 540]. As was seen with Clower's 1959 article, there is a recognition that when dealing with less than perfect competition the equilibrium model builder must come to terms with how imperfect competitors deal with the incomplete information provided by the market. Richardson's 1959 article goes further to consider what we would have to assume to achieve an equilibrium in a general equilibrium model. In the case of the firms in a model of general equilibrium, he argues that the model builder must explicitly address the firms' knowledge – but most important, he notes like Clower that firms cannot know everything they need to know in real time to *guarantee* reaching an equilibrium. Of course, some knowledge can be assumed to be available. Specifically, every decision maker can be assumed to have knowledge of what Richardson called *primary* conditions (e.g., private knowledge of one's own utility function or production function). But the firm needs to know more to be able to make decisions about quantity or price, particularly when the general equilibrium model includes a market for investment goods. But if we leave out the question of an investment market and just consider the dynamic process of reaching a general equilibrium, each decision maker must also have knowledge of what Richardson calls *secondary* conditions. His secondary conditions involve knowledge about the activities of other market participants.[1] That is, as with Clower's ignorant monopolist, it is one thing to know the last market period's equilibrium price; it is quite another to know about the next market equilibrium price, since one would have to know everyone's demands or supplies in the next period. Of course, such secondary knowledge is usually not directly available and thus the individual must form what Richardson called 'rational expectations' if any equilibrium is to be attained.[2] And, like Arrow, he too noted that some form of imperfect competition needs to be included if we are to explain the dynamic process of achieving an equilibrium.

---

1. Note that he says [p. 225]: 'This terminology is meant to convey no difference in the importance of these two kinds of conditions; a distinction is made between them because of the different ways in which information about them has to be obtained'. Eventually, he calls these 'primary information' and 'secondary information' [e.g., p. 229].

2. This has little to do with the rational expectations assumption introduced by John Muth [1961]. And unlike typical model builders today, Richardson is not using 'rational' to mean maximizing. Instead, he merely means that it is possible to deduce the expectations from a set of assumptions that may or may not be influenced by observations of past events.

## 4.2. RICHARDSON ON COMPLETING
## AN EQUILIBRIUM MODEL

Richardson focused on what it takes to complete a general equilibrium model of a whole economy and his main focus was on the information needed by each autonomous individual participating in the economy. But he warned [p. 223], 'that the familiar "general equilibrium of production and exchange" cannot be properly regarded as a configuration towards which a hypothetical perfectly competitive economy would gravitate or at which it would remain at rest'. Like both Arrow and Clower, to complete such a model he narrows his focus to identifying what must take place prior to attaining the general equilibrium.

### 4.2.1. The informational requirements for a perfectly competitive equilibrium

Like most critics of relying only on Marshall's equilibrium models to be the basis for explaining some aspect of economics, Richardson lays out all of the necessary conditions for the existence and maintenance of a state of general equilibrium.[3] He says that a sufficient condition for a whole group or an economy to be in equilibrium is for every member of the group or economy to be in his or her personal equilibria.

Richardson thinks we build equilibrium models in economics so that we can say what autonomous agents in an economy will choose to do under 'certain postulated conditions'. Of course, there is no reason to think there are direct connections 'between objective conditions and purposive activity'; instead, what matters is the 'immediate relationship between *beliefs* about relevant conditions and *planned* activities which may or may not prove possible to implement'. As he says, what the neoclassical economists' equilibrium model's postulated autonomous agent 'needs to know are the conditions of the allocation problem as laid down by the model builder' [1959, p. 224]. If the agents have such knowledge, they will engage in the postulated equilibrium activities – moreover, such equilibrium activity will maintain the equilibrium so long as there is no change in the 'determining conditions'. However, he says those informational requirements will apply even for simple equilibrium models.

Richardson recognizes that not all models of individual agents are simple. And he notes, while full information would be a logically sufficient condition for attaining the postulated equilibrium state, it is not a logically necessary condition. This, of course, is evident in Clower's model of the ignorant

---

3. See Richardson [1959, pp. 223–24].

monopolist who lacked full information yet attained a state of equilibrium. As we saw in Chapter 3, all that was required was that the monopolist believed its chosen activity was optimal; moreover, nothing Clower's ignorant monopolist does provides information that is inconsistent with or would cause it to doubt its assumption concerning the demand curve. But as Richardson would have pointed out, economists have too often concentrated only on states of equilibrium in which the beliefs (or assumptions in Clower's model) are actually true.[4] What concerns Richardson is that we can never 'identify equilibrium activities without reference both to objective conditions and to states of belief' [p. 224, fn1]).

### 4.2.2. The consequences of reaching a general equilibrium

Given all this about informational requirements, Richardson goes on by adopting the traditional assumption that the equilibrium for a group (or for a system as a whole) is reached when all of its autonomous individual agents are in equilibrium. This in turn means that every individual believes his or her projected activities are optimal. Moreover, implementation of the projected activities goes according to the projected plan and thus there is no reason to alter their planned activities. And if the planned activities are to be carried out as intended – as in the case of the Clower's monopolist in his Model III – assumptions (or expectations) must not be contradicted by subsequent events. Of course, if the assumptions (or estimates) are correct, then the equilibrium is guaranteed. And Richardson said that for this to be the case, the agent must have adequate primary knowledge (e.g., about its own preferences and the usual technical matters) as well as adequate secondary knowledge about the activities of other agents in the group or system.[5]

With this in mind, Richardson proceeds to examine the usual idea that identifies any general equilibrium of production and exchange with what advanced textbooks of the 1950s and 1960s would consider perfect competition – a system characterized by the assumption of very many buyers and sellers of homogenous goods, and in which there is 'perfect mobility', that is, there are no restraints on the activities of the buyers and sellers, such that all potential gains can be pursued. Builders of equilibrium models of perfect competition usually presume that all buyers and sellers have all of the necessary

4. However, we saw in Arrow's view of an imperfect competitor and Clower's Model II (as well as Clower's 1955 article), when information is less than adequate, one can assume that eventually the required full information is still obtained by trial and error.
5. He also points out that, from the point of view of an observer external to the system (such as the model builder or someone using the model), this might mean that the activities of all of the agents in the group or system being modelled must be compatible with each other as well as their 'primary conditions' [p. 225].

information, but Richardson observes that 'this requirement is usually left obscure, and its relationship with the other conditions of the model is not examined' [p. 225]. Moreover, the usual builders of equilibrium models presume the equilibrium state can be 'identified and made to depend on external conditions represented by the preferences of individuals, the endowment and original distribution of resources, and the state of technique' [pp. 225–26].

Going on [p. 226], he identifies the familiar two special properties that any equilibrium has. First, every utility-maximizing consumer spends their whole income and every profit-maximizing producer is 'producing as efficiently as possible'. And for the whole economy, all scarce resources are fully employed and their sale is the source of the individuals' incomes 'as determined by the distribution of skill and property originally assumed'. Among the more mathematically motivated model builders of the 1950s and 1960s, much of the discussion of such a model would be concerned with proving the existence and stability of any claimed equilibrium solution to such a model. Richardson deliberately avoided such a discussion of existence and stability for equilibrium models.[6] Instead, he just focuses on the characterization of a perfectly competitive economy in an alleged equilibrium state. However, he still stressed that such an equilibrium state was dependent on the particular set of beliefs of the participating agents being modelled. Richardson goes on to say that the model builder needs to identify what these beliefs have to be so that the equilibrium economy 'can be regarded as self-perpetuating' – doing so includes the belief that 'their incomes and the prices of commodities were regarded as fixed and given' and since everyone is maximizing, nobody wishes to change anything.

The second set of special properties are about the consistency and compatibility of the various activities that would guarantee that all plans can be carried out and that all beliefs or expectations that agents based their plans on will be shown to be true or correct.[7] On this basis then, all markets are seen to be cleared as everyone's planned purchases and sales are fulfilled. And so, it must be the case that the beliefs required for the attainment of the equilibrium state must have been established. This also includes the primary information and the belief that current prices will be permanent so long as there are no changes in the given situation.[8] He next goes on to consider 'whether

6. And I will postpone a discussion of these necessary properties of an equilibrium model until Chapter 5.
7. Richardson [p. 227, fn1] notes that 'there are in fact an indefinite number of expectational conditions which would, under reasonable assumptions as to human motivation, secure the maintenance of the general equilibrium. And though some of them no doubt appear less unrealistic than others, there seems no clear reason why any one set is most appropriate'.
8. He does note that this might 'untidily' require that no producers wish to increase their output.

the conditions of belief which will ensure the maintenance of equilibrium activities will also ensure their adoption' [pp. 227–28]. And to address this question, Richardson turns to consider a disequilibrium state that might exist prior to the attainment of an equilibrium.

### 4.2.3. Considering the disequilibrium before reaching the equilibrium

For this consideration, Richardson asks us to assume that prices, production outputs, etc. are not equilibrium values but also that everyone, nevertheless, expects equilibrium outputs and prices will be attained in the future. And then he asks, 'Will the system move to equilibrium?' [p. 227]. To answer this question, Richardson says that we need to distinguish between the usual assumption that 'entrepreneurial skill is undifferentiated' and the assumption that 'no entrepreneurs are alike'. He first considers the usual assumption that all are alike.

This first assumption implies that actions of any entrepreneur are completely indeterminate. Expected product and factor prices are such that profits are normal for every product being considered for production and so there is no reason to favour one over the other. And in this case, we have no reason why any particular equilibrium would be attained. But, once the model builder instead assumes that no entrepreneurs are alike, it would mean that some have a comparative advantage over others – the allocation of production outputs can be determined as each producer will have one particular niche where its profit will be maximum. And in this case, Richardson says equilibrium prices will be expected such that all producers will produce where the equilibrium will be attained and expectations will be fulfilled.

Richardson then turns to consider what beliefs are now required to guarantee that agents will engage in the equilibrium activities. He asks whether it is enough just to assume that entrepreneurs believe that equilibrium prices will be maintained? Should current prices happen to deviate from the equilibrium prices, any expectation of their continuance will clearly be disappointed and thus any plans based on current prices cannot be fulfilled. In such a model as this, the equilibrium can be attained only if expectations are revised somehow. The subsequent plans that would be based on the new expectations will need eventually to create prices that will be projected into the future to serve as a basis for new plans and the back and forth continues. If this behaviour is included in the equilibrium model then the model will exhibit the needed movement to a general equilibrium. But, Richardson warns, there is a fundamental objection to this kind of model: 'are we entitled . . . to conjure up expectations as we please?' [pp. 228–29].

He goes on to note that expectations depend on available information and that availability depends on the economic system we have chosen to assume. The expectations cannot be viewed as a completely extraneous element of our model for which we do not have to account for in our model building. Instead, he thinks [p. 229] that since we usually 'assume that economic agents generally act rationally on the basis of their beliefs; we ought similarly to assume that beliefs or expectations themselves are rational' when the beliefs are augmented by 'adequate evidence or information'.

Again, Richardson here is discussing building equilibrium models of group or system activity. The availability of what he calls primary knowledge is not dependent on what we assume about the nature of the system or organization and as such it can be assumed that all entrepreneurs have knowledge of the current so-called 'state of the arts'. But, this is not so for what he calls secondary knowledge. The availability of information needed for the secondary knowledge depends on which system we have postulated.

### 4.2.4. The availability of needed information in a competitive general equilibrium model

Equilibrium model builders, of course, can assume consumers are fully informed, not only about their preferences but also about the nature of all the goods they might buy. Doing so, however, would not be realistic and would lead to a model that does not perfectly explain the working of the system or economy. It might be acceptable for an idealized system or economy that some might find worth considering – model builders could assume that producers have some of what he calls secondary information about the projected activities of other producers in their group, but then they would have to postulate a system that would provide such information. For example, a model builder might assume the members of the group plan their respective activities together. The model builder also could go much further and have them turn themselves over to some sort of group planning authority.[9] If they did this, there would be no need for secondary knowledge. But of course, adopting this means of avoiding how to explain the acquisition of the required secondary information needed for the attainment of an equilibrium would be inconsistent with the reason for building a model of general equilibrium. If equilibrium model builders choose to avoid such an inconsistency, Richardson asks, how do the producers in their models obtain sufficient secondary information to

9. In Canada we have had such Marketing Boards for agricultural products such as milk and eggs.

be able to assess the profitability of their own investments? Needless to say by now, this is a fundamental problem for building any general equilibrium model. And it is one that the textbooks' theory of a perfectly competitive economy is unable to give an acceptable solution. Richardson says that the problem arises simply because the perfectly competitive model presumes the members of the group or system engage in their activities independently but he thinks equilibrium model builders need to recognize that primary knowledge alone is not enough.

In any perfectly competitive industry in which there are very many producers, any opportunity available to one is available to all. Thus each producer cannot decide on an optimal future output level or investment without some knowledge of what the other producers are going to do.[10] And, Richardson goes on to say, such a mutual interdependence is a barrier to obtaining the needed secondary information. And if one is going to build an equilibrium model of such a situation, one still must explain in one's model how this barrier can be overcome. Surely, with any consideration of imperfect competitors and in particular, oligopolies,[11] this problem would have been addressed.[12] One might think that, unlike an oligopoly, a perfectly competitive producer might have a determinate equilibrium solution since the perfect competitor cannot affect the given price. But, as Richardson points out, it is the unknown expected future price that is the relevant price. Current prices alone can only provide inadequate information if this is about the disequilibrium process of reaching an equilibrium and not just about what takes place at an equilibrium. And as he says [p. 231], 'de-centralised systems can work only provided their constituent members can obtain the minimum necessary secondary information' and so equilibrium models must explain how such secondary information is obtained.

While such information may be necessary for deliberate activity, it in no way guarantees that any particular action will be taken and especially whether the action is to the benefit of society as a whole. If the equilibrium model builders, as usual, are going to assume that individual activities are not deliberately coordinated, then Richardson says two essential requirements still must be met. The first requirement is fundamental even though, according to

10. Again, in Canada during the sharp rise in oil prices in the mid-1970s, many new wells were invested in on the presumption that few other wells would be drilled. Of course, such a presumption contributed to an eventual fall in oil prices due to a resulting excess supply.

11. For readers unfamiliar with this jargon, it refers to a market with very few producers but many demanders.

12. Richardson says that this problem has often been noticed but usually dismissed as some minor 'logical catch' [p. 231]. But he thinks this is curious given all the attention in the 1950s to the models of an oligopoly and their alleged indeterminate equilibria.

Richardson, models of a competitive economy fail to meet it. It is that their equilibrium model must include an explanation of the availability of opportunities and of how those opportunities are known to the agents in the model. And the second requirement is that the model needs to show that these opportunities are such that they will be taken given the motivations that are assumed in the model and that they lead to actions that are, in effect, of the social interest of those agents recognized in the model.

Most importantly, the textbooks' theory of perfect competition fails to explain 'how the supply of secondary information is functionally related to the structure of the system assumed' [p. 232]. At this point in his article, Richardson introduces three ways producers could obtain the necessary secondary information. The most obvious is 'explicit collusion'. Less obvious is his second way, which he calls 'implicit collusion'. The third way is what he calls 'restraints'. About the third way, he is a bit vague but he thinks there is no reason why producers would not at least know what the maximum possible competitive supply can be in some relevant future time if there are somehow restraints, perhaps natural, that limit that maximum. Similarly, the minimum can be estimated based on 'the momentum of previous commitments' [p. 234]. The second way is most interesting if one thinks an equilibrium has been established and everyone thinks the equilibrium prices will remain unchanged. In effect, as we might now say, 'if it ain't broke, don't fix it'. While implicit collusions might be a way of explaining the maintenance of an equilibrium state, it can hardly be the basis for explaining how that equilibrium is reached. With all of this, Richardson concludes [p. 233] 'that the beliefs required to maintain the perfect competition equilibrium could be rational', by which he means they are consistent with available information, but this may only be 'under very special conditions and in a very special way'. However, while such beliefs might be easy to maintain in a state of equilibrium, 'we have as yet discovered no way in which entrepreneurs could obtain any information about each other's projected activities and therefore no way in which rational expectations could be formed'.

The bottom line of Richardson's article is simply that once an equilibrium model builder recognizes the need to address the matter of informational requirements for reaching and maintaining the equilibrium being modelled, it becomes difficult to see how assuming prefect competition could ever provide the circumstances leading to the needed information or knowledge. While it might be mathematically possible to build a model of an economy in the state of equilibrium, it is not possible to maintain the usual assumptions needed for a perfectly competitive economy to reach that state of equilibrium if one starts the consideration at a state of disequilibrium. During the time it takes to reach the model's state of equilibrium, as in the case of Arrow's 1959 article, the model's agents would have to engage in activities that are denied by the perfectly competitive model.

Except for Richardson's efforts to identify what is needed to overcome the failure of a perfectly competitive equilibrium model to fulfil the informational requirements, the logic of this situation was exactly that which Arrow presented with his call for an explanation of the dynamics of price adjustment to a state of market equilibrium. As we saw in Chapter 2, if we start by considering the state of market disequilibrium, the going market price is not sufficient information to guarantee the eventual attainment of the market equilibrium. Not only must the actual adjustment process be explained if we are going to produce a complete model of the market equilibrium, Richardson is arguing that requirement is also true for models of general equilibrium.

# CHAPTER 5

୦ᐯ୦

# Equilibrium concepts and critiques

## *Two cultures*

It is not always clear when economists argue over the notion of equilibrium whether they are really talking about the same thing. The nature and cause of this possible disagreement will be examined in this chapter. But let us first consider what a prominent historian of economic thought thinks about the modern economists' use of the notion of equilibrium – consider Roger Backhouse's observation [2004, p. 291]:

> Arguing in terms of equilibrium would appear to be an inescapable feature of the present-day economic orthodoxy. Economists argue about which is the appropriate equilibrium concept to use in a specific situation but the necessity of arguing in terms of equilibrium is taken for granted – so much so that the question of its desirability is not even raised.

It is not clear that its desirability has been unquestioned. Even one of the early mathematical equilibrium model builders, Abraham Wald, warned about relying on mathematical equilibrium models [1936/51, p. 368]:

> [C]onclusions have often been drawn from mathematical formulas, which, strictly speaking, are not conclusions at all and which at best are valid only under restrictive assumptions. The latter may not have been formulated, not to mention efforts to discover to what extent these further assumptions are fulfilled in the real world.

Critics of relying just on the equilibrium nature of an equilibrium model might go even further. For example, Nicholas Kaldor long ago suggested [1934, p. 70]:

> For the function which lends uniqueness and determinateness to the firm – the ability to adjust, to co-ordinate – is an *essentially dynamic function*; it is only required so long as adjustments are required; and the extent to which it is required ... depends on the frequency and the magnitude of the adjustments to be undertaken. It is essentially a feature not of 'equilibrium' but of 'disequilibrium' ...

Other critics have gone still further – for example, Robinson critically observes [1953–54, p. 85]:

> The neoclassical economist thinks of a position of equilibrium as a position towards which an economy is tending to move as time goes by. But it is impossible for a system to get into a position of equilibrium, for the very nature of equilibrium is that the system is already in it, and has been in it for a certain length of time.

Unfortunately, as we will see in this chapter, many of the critics do not always understand modern-day equilibrium model builders, particularly those model builders who continue to see no problems building elaborate mathematical equilibrium models other than those problems involved in matters of mathematical adequacy. And most of these modern-day model builders do not seem to find the views of the critics to be of any interest.

In the previous chapter I mentioned that Richardson focused on what would have to take place for an economy to move from a state of disequilibrium to a state of equilibrium. Interestingly, in a footnote[1] he says he explicitly chose to avoid discussing what builders of formal equilibrium models during the 1950s and later in the 1960s were calling the problems of existence and stability. These usually are the problems of specifying the assumptions of the formal general equilibrium model – first to be able to prove the existence of a solution to a model of Walrasian general equilibrium and second to be able mathematically to prove the so-called stability of that solution. A solution for such a model is a set of equilibrium prices, one for each market whereby all markets are clearing simultaneously and thus all market participants are maximizing. The stability problem is about mathematically proving that markets are stable in the sense that for any slight deviation from the equilibrium state, the market will self-correct and return to an equilibrium.[2] These problems as

---

1. See Richardson [1959, p. 226, fn1].
2. In Chapter 2, the discussion was about Arrow's specification of his equations [2.4] and [2.6] and his inequality [2.5] which together was his solution to the stability

well as the so-called uniqueness problem[3] need to be solved if one is going to use the formal model to explain all market prices or all market sales.

Parenthetically, it should be pointed out that what is now called the stability problem is not in any way related to the old hog cycle or cobweb stability question.[4] In the idea of the cobweb – where there is a sequence of alternating quantity and price adjustments – stability depends on whether the relative slopes of the demand and supply curves assure convergence. Roughly put, when the slope of the demand curve is steeper than the slope of the supply curve, the sequence of price and quantity adjustments can cause the price to move away from the equilibrium. When the relative slopes are the other way, the price and quantity adjustments can cause the price to converge to the equilibrium or clearance price. In today's equilibrium models, it is not usually presumed that there is a sequence of alternating price and quantity adjustments and instead price and quantity are determined simultaneously. Today, stability problems are nevertheless solved by placing restrictions on the relative slopes of the demand curve and the supply curve. However, as will be explained in Chapter 7, some solutions involve invoking a form of collusion since it makes the behaviour of demanders and suppliers interdependent.

It is important to notice that today's perspective on the stability problem and equilibrium model building is rather narrow in that it does not seem to address the concerns of critics of equilibrium model building from the 1930s and 1940s such as those of Joan Robinson, Nicholas Kaldor, Ludwig Lachmann, Fritz Machlup, George Shackle, or even of Friedrich Hayek, or the more recent critics such as Tony Lawson and Jànos Kornai as well as most of those critics who currently work in the area of evolutionary economics.[5] When it comes to talking about such things as determinateness, existence, uniqueness, or just stability, is everyone talking about the same things? According to E. Roy Weintraub [2005], they definitely are not.

## 5.1. TWO CULTURES

As I noted in Chapter 1, Marshall saw the notion of an equilibrium to be central in his method of explanation. About the notion of an equilibrium he says, 'When the demand price is equal to the supply price, the amount produced has no tendency either to be increased or to be diminished; it is in equilibrium'

---

problem for a simple market equilibrium. His specification was necessary for the proof of a stable market represented by his equations [2.1] to [2.3].

3. Which involves specifying that model's assumptions to be able to prove the absence of multiple solutions as will be explained below.

4. This matter has a long history. For example, see Ezekiel [1938].

5. I will be discussing evolutionary economics in Chapter 11.

[1920/64, pp. 287–8]. Many years later Machlup also considered how we look at the notion of an equilibrium in economics. As he said [1958, p. 9]:

> [W]e may define equilibrium, in economic analysis, as a constellation of selected interrelated variables so adjusted to one another that no inherent tendency to change prevails in the model which they constitute. The model as well as its equilibria are, of course, mental constructions (based on abstraction and invention).

But a consistent critic of equilibrium explanations, Robinson, observed, 'Long-period equilibrium is not at some date in the future; it is an imaginary state of affairs in which there are no incompatibilities in the existing situation, here and now' [1962, p. 690]. Another critic, John Henry adds, 'it is just this point of relevance and realism which raises doubts about the credibility of equilibrium theory in explaining social phenomena and, thus, in its ability to assist in the formulation of policy' [1983–84, p. 217].

As will be discussed in this chapter, these economists seem to be discussing something different from that what can be found in the textbooks that I used in my graduate program. For example, consider the view of Robert Dorfman, Paul Samuelson and Robert Solow [1958, p. 351]:

> [W]e can't blithely attribute properties of the real world to an abstract model. It is the model we are analyzing, not the world. . . . One test is provided by the existence problem. Just because no real existence problem can occur, a system of equations whose assumptions do not guarantee the existence of a solution may fail to be a useful idealization of reality.

In this regard, Roy Weintraub observed, 'If mathematics and economic theory are interwoven, then to understand the practice of economics we must understand the practice of mathematics' [1985, p. 171]. But, the famous historian of economics, Joseph Schumpeter, went further to observe, 'One sometimes has the impression that there are only two groups of economists: those who do not understand a difference equation; and those who understand nothing else' [1954, p. 1168].

In a 2005 article about Tony Lawson's concept of an equilibrium, Weintraub argues that there are not just two groups of economists, but in effect two cultures when it comes to equilibrium concepts and equilibrium model building – and they rarely seem to be talking about the same thing. The first includes those who learned about the concept of equilibrium many decades ago and usually in the historic Marshallian tradition, which, according to Weintraub, was that, 'Equilibrium was . . . a manifestation of a series of arguments that had physical meaning' [p. 451]. For this older culture, equilibrium is a simple matter concerning empirical realism. The newer culture includes those like Paul Samuelson and his colleagues who became active in mathematically examining the logic of the old arguments based on various notions of equilibrium. The main tool of this newer culture has always been the formal or mathematical general equilibrium model and so problems of stability or existence were always seen to be about properties of formal models rather than as Weintraub notes, having a Marshall-inspired physical meaning such as 'an egg in a bowl'. So, between these two cultures we have two different views[6] of what the idea of a state of equilibrium is about.[7]

The two equilibrium cultures became separated soon after 1940 – and by the time I started my PhD in the 1960s, there were organized efforts by the newer culture to train us to be 'quantitative economists' and for this purpose, in my case in the U.S., my PhD was financed with federally provided money.[8] Most of our training was explicitly involved with questions of proving existence, stability, uniqueness, etc. For us, equilibrium was always an explicit property of a formal model.

In Boland [2014] I also discussed a related issue of a dominant culture, namely, that concerning the culture found in mathematics departments and one that has overtaken most modern economics departments. This culture goes beyond just saying equilibrium stability is a mathematical property of a model. As I explained there [p. 247],

> [T]he culture of the mathematician is that when building a model one's objective is to construct a proof and to do so *by any means whatsoever*. The realism of one's assumptions is not important; what is important is whether the proof is logically sufficient and then – and most important – whether one's proof is deemed *elegant*.

Clearly today, the task for the critics is to convince today's formal model builders that realism must matter. But, I think it is not just a simple matter of the realism of a model's assumptions as it is more a matter of what the

6. Note that what these two views mean by 'equilibrium' should not be confused with the two meanings of 'model' which I briefly mentioned in the Preface and discussed in detail in Boland [2014]. In that discussion I identified two views of what one means by *a model* that differ depending roughly on whether one got their economics PhD before or after 1980. Those before 1980 view a model as a representation of a previous theory or hypothesis. Those after 1980 view a model and theory interchangeably as both are presumed to be expressed using formal mathematics.
7. While he was not as explicit, a couple years earlier Alan Kirman [2003] made a similar distinction between different viewpoints on states of equilibrium. For Kirman, the difference is between what he called the classical view and the neoclassical view. He was making this distinction without making as strong a point as Weintraub did about the formalist model builders and their narrow interest in seeing equilibrium as nothing more than a property of a model.
8. As I explained in the Preface, PhDs like mine were financed with lucrative and tax-free NDEA fellowships and with the financing of professorships to train us in high-tech mathematical model building.

concept of an equilibrium meant for the older culture in the physical sense regardless of any worrying over the realism of a formal model's assumptions.

## 5.2. EQUILIBRIUM CONCEPTS INVOLVING TIME, DYNAMICS AND PROCESS

Hicks is usually recognized as a major early promoter of building Walrasian general equilibrium models in economics. He has said, 'it is quite true that we assume the economic system to be always in equilibrium. Nor is it unreasonable to do so' [1939/46, p. 131]. For the 'economic system' to be in equilibrium constitutes what we today call general equilibrium. While it might be easy for some model builders to agree with Hicks, a more cautious view is taken by Yves Balasko [2007, p. 413]:

> General equilibrium theory shortcuts the study of the out-of-equilibrium price dynamics by jumping to the stationary points. In addition to simplifying the mathematics, this shortcut dispenses economists with getting into the specifics of that dynamics. But many questions cannot be answered without some understanding of the out-of-equilibrium price dynamics.

But the critic Robinson would still complain, 'The hard core of logical analysis in [Marshall's] Principles is purely static – it applies to an economy in which accumulation has come to an end' [1978, p. 132]. Criticism of static equilibrium models is not new. For example, Arthur Smithies noted [1942, p. 26]:

> [T]he equilibrium theorists, in their enthusiasm for existence theorems, have too frequently ignored the equally important problem of whether a possible equilibrium will in fact be approached by an economy starting from an arbitrary disequilibrium position. The process theorists on the other hand have been too prone to overlook the possibility that the process which they analyze may eventually result in equilibrium.

As I suggested in the discussion of Arrow's 1959 article, it is one thing to demand that equilibrium model builders recognize the need for dynamics, it is another to claim that there is also the need to explain how long it takes to reach the model's equilibrium. But as I noted in Chapter 1, textbooks usually rely on Marshall's mode of explanation with its reliance on states of equilibrium having been reached. Yet, as Mohammad Dore [1984–85, p. 196] critically observes about Marshall's equilibrium mode of explanation,

> The longer it takes prices to converge to equilibrium the longer [Marshall's] *ceteris paribus* assumption must hold. Consequently convergence as time approaches infinity is logically inconsistent with the *ceteris paribus* assumption.

It is likely that all of the critics of neoclassical equilibrium models who claim such models are unrealistic because they involve a 'static' form of equilibrium[9] do not understand the inherently dynamic nature of the concept of an equilibrium as opposed to the static nature of a balance.[10] But as the 1959 articles by Arrow and Richardson demonstrated, the problem with such neoclassical equilibrium models is not that they are inherently static – the problem is that they are incomplete. A complete equilibrium model is inherently dynamic.

Perhaps, if the members of the two cultures would spend less time complaining and more time recognizing that the other culture has a point or two worth addressing, then progress could be made toward building complete equilibrium models and thereby closing the gap. Surely, a complete model must not only recognize dynamics, it must include explicit behavioural adjustment-process assumptions. The sticking point will always be that of the realism or at least the plausibility of those behavioural assumptions. Failure include explicit behavioural adjustment-process assumptions will always mean that formal equilibrium models will remain limited and problematic whenever they are used to create business or governmental economic policies.

The critics need to change their approach, too. At a minimum, they need to try understanding what is of concern to formal economic model builders and why. Of course, it would also help if the formal model builders would stop dismissing the concerns of the critics and instead at least try to understand them. In the next two sections, I will try to assist both cultures in such efforts.

## 5.3. EQUILIBRIUM CONCERNS OF THE FORMAL MODEL BUILDERS

Today's modern economics departments, as I noted in Boland [2014] and briefly quoted above, are dominated by the culture of the mathematics departments that values logical sophistication and rigour more than the question of realism. And, of course, many modern economics departments today include researchers of this culture that devote their research to building formal equilibrium models. It could be argued that progress in economic model building will occur – perhaps only occur – when the culture of the mathematics department is abandoned or at least suppressed. However, giving up that culture of

9. For examples, see Robinson [1974] or D'Agata [2006].
10. It does not seem to bother formal economic equilibrium model builders that the concepts of balance and equilibrium are often confused. Even physics textbooks often use the words interchangeably but physics textbooks are at least clear about the dynamic nature of any equilibrium. To a certain extent, Arrow was addressing this confusion in his 1959 article.

the mathematics departments does not require giving up the culture of the formal economic equilibrium model builders.

From the beginning of efforts to use formal Walrasian general equilibrium models to explain an economy, builders of those models recognized that they must mathematically prove the provision of two important aspects of their equilibrium models if their models are to be used for explanations of such things as prices for all goods. In this regard, Dorfman, Samuelson and Solow [1958, p. 349] are explicit:

> [I]n connection with abstract Walrasian systems, the main question that seems to have been studied in the literature has to do with the *existence* of an equilibrium solution to the collection of equations and with the *uniqueness* of the equilibrium if it exists.

In the remainder of this section I will discuss the standard concerns of formal economic equilibrium model builders to see if it is possible for those model builders to address those concerns without surrendering to the culture of the mathematics departments. In this regard, it is still important to keep in mind that much of the early formal model building during the twentieth century was devoted to examining the logical adequacy of existing non-formal theoretical models that claimed to provide explanations of various economic phenomena such as prices, sales, or resource allocation using the idea of a state of equilibrium.

### 5.3.1. Existence

Formal economic models such as those that purport to explain, say market prices, usually consist of many assumptions represented in the form of equations. The model is 'built up' one assumption at a time. While it should be unlikely, it is possible to end up with a set of assumptions which is not consistent in that a later assumption might logically contradict some earlier assumption. This is not so easy to see if the model consists of very many assumptions. If we wish our explanation to be logically valid, of course, it cannot allow for such an inconsistency. And, even when there are multiple solutions to a model – that is, when we can deduce from the same set of assumptions one or more sets of prices that allow all markets to clear – our formal model is at least not inconsistent. From the beginning, the primary means of proving consistency has been to prove the existence of a solution to the system of equations. And as Weintraub observes, 'The theorem proving the existence of general equilibrium in a competitive economy, which necessarily involved specifying the conditions under which such an equilibrium would exist, is an extraordinary achievement of twentieth-century economics' [2011, p. 199].

Unfortunately today there is little discussion of a model's consistency. I suspect that this is a consequence of the dominance of the mathematics culture that is less concerned with the explanatory nature of economic models and is more concerned with techniques of mathematical proof.

### 5.3.2. Uniqueness

From the time of Walras, general equilibrium models were built to provide an explanation of potentially observed prices. If one does consider such an explanatory purpose of economic models, then uniqueness needs to be considered. This is because, as I said, consistency alone does not eliminate the possibility of multiple equilibria. But, if your explanation allows for more than one set of equilibrium prices, your explanation would be incomplete as you would still have to explain why one set of values for the prices is observed but the other logically possible sets allowed by the explanation are not observed. A complete explanation would not only have to explain why prices are what they are but also why they are not what they are not.[11] Multiple equilibria leave unexplained why one or more logically possible sets of prices are not observed. A proof of uniqueness is thus a key aspect of a proof of the completeness of a model's *explanation*.[12] But as Alan Kirman observes, 'uniqueness of equilibrium can only be bought at the price of extreme assumptions on individuals' [2003, p. 472].

### 5.3.3. Stability

As we see in an early textbook for formal equilibrium model building, 'If we add some dynamical assumptions describing the response of price and quantities to disequilibrium situations, then we can also study the *stability* of possible equilibria' [Dorfman, Samuelson and Solow, 1958, p. 349]. While it might be easy to see the questions of existence and uniqueness as only matters of mathematical elegance or good form, an equilibrium model builder's concern for stability is more than that. As Fisher has observed, 'If equilibrium analysis

---

11. See Nikaido [1960/70, p. 268].

12. Some mathematicians might view a model that explains a set of endogenous variables as a mapping from a single set of values of its exogenous variables to a single set of values of those endogenous variables. To the extent to which such a model is an explanation, the mapping must be 'well defined', which is roughly another name for being unique. So, at worse, the usual requirement for a proof of the uniqueness of the equilibrium can be just a formal mathematical requirement having nothing explicitly to do with the simple logical concern for a complete explanation. For more about this logical concern, see [Boland 1989, pp. 123–25].

is to be justified, the crucial question that must first be answered is one of stability' [2003, p. 76]. As I noted in Chapter 2, Arrow [1959] argued that a formal model of a market's equilibrium is not complete unless it also includes equations about dynamics that provide the minimum necessary price adjustment equations and relational conditions to assure that the adjustment is in the correct direction. That seems somewhat contrary to what Dorfman and his colleagues were suggesting in the above quotation (viz, 'If we add . . .') which some may read as saying that it is an optional consideration.[13]

## 5.4. CONCERNS, BEYOND REALISM, OF THE CRITICS OF FORMAL EQUILIBRIUM MODELS

As I have repeatedly noted, the main concern of the critics of equilibrium-based explanations or models is that there seems to be no concern for realism, nor, as the formal equilibrium model builder Wald even noted, any concern for the extent that the models assumptions 'are fulfilled in the real world'. But the realism of the explicit assumptions of equilibrium-based models is not the only complaint. Many critics go further to focus on what is missing in the usual Walrasian general equilibrium model and those will be the next topics of discussion.

### 5.4.1. Knowledge and information

Some of the economists who believe that markets can solve all social problems also believe market equilibrium prices provide sufficient information to assure that a socially desirable equilibrium will be attained and retained.[14] But as I explained in Chapter 4, once model builders start considering how that equilibrium and its prices were reached in the first place, attaining the equilibrium can be seen to require more information than just previous equilibrium prices. This is an important consideration for equilibrium model building in economics so I will return to the question of the informativeness of equilibrium and disequilibrium prices later in Chapter 8.

### 5.4.2. Expectations

Kirman observes that 'In any intertemporal situation with uncertainty, agents must form expectations or at least have beliefs' [2003, p. 479]. In Chapter 4

---

13. I suspect they meant something like 'Whenever' rather than a subjunctive-like phase starting with 'If'.
14. For example, see Hayek [1945].

I talked about Richardson's argument that once we start considering the dynamic process that might lead to an equilibrium, ignorance of the future state of affairs requires forming expectations in order to decide what to do next. While it is tempting to think you need only look at available information to learn what to expect – and thereby form so-called rational expectations – for the expectations to be accurate would require an inductive logic. But, as I briefly noted in Chapter 1 (fn. 22), there is no such valid form of logic.[15] Moreover, as Keynes [1937] explained, expectations are always guesses however informed.

### 5.4.3. Uncertainty

Whenever one is recognizing the need to deal with expectations one is also recognizing this need as a means to deal with uncertainty. However, it is important to keep in mind that uncertainty is not the same as risk. I say this because once we say 'uncertainty', the usual response is that we can measure the degree of uncertainty of a future event of interest with a probability. But, the uncertainty that the critics think equilibrium models must recognize cannot honestly be represented with a probability. As Paul Davidson [1991] explained – and both Frank Knight and Keynes stressed – risk is about situations where you know the probabilities (e.g., roulette wheels), but uncertainty is when you do not have such knowledge and hence no basis for assigning a probability.[16]

### 5.4.4. Increasing returns to scale

Increasing returns to scale might not seem to be an obvious concern but as Hahn observes [1981, p. 125]:

> An equilibrium of the economy is a state in which the independently taken decisions of households and firms are compatible. . . . The equilibrium prices impose order on potential chaos. To show that these equilibrium prices can indeed be found, one needs a number of further assumptions. The most important of these is the absence of significant economies of scale in production.

15. I will discuss this issue more in Chapter 6. For a fuller discussion, see Boland [2003, ch. 1] or [2014, ch. 11].
16. While some economists would see this as a matter worthy of subjective probability as a measure of ones opinion of the future event, there are many situations in which the future event in question is singular. Singular events have a probability of one or zero – they either happen or they do not. A probability of say 30% makes no sense in the case of a singular event, only a zero or one does.

At base, those critics who think general equilibrium models can be realistic say model builders should recognize that a state of equilibrium must involve some form of imperfect competition. And to be realistic general equilibrium models must also involve competitive firms producing at a level of supply where profit is zero. At that level, their average revenue and average costs are equal since competition itself eventually causes total profit to be zero – but for an imperfect competitor this will be so only at a point for which the downward sloping demand curve facing the firm is tangent to the downward sloping segment of the firm's average cost curve.[17] That is, an imperfect competitor will be producing at a point on the average cost curves where there is increasing returns to scale (or as it used to be said, there are 'economies of scale'). Such a point will be to the left of the producer's conceivable perfectly competitive equilibrium (i.e., where average cost is at its minimum).[18] So, to stress, whenever average costs are falling with the supply output, the imperfectly competitive suppler in long-run competitive equilibrium will necessarily be facing increasing returns to scale.[19] While this may be true for matters of realism, this is not why Hahn in the above quotation from his 1981 article was discussing 'economies of scale' (i.e., increasing returns to scale). His concern is that the mathematics of solving for an equilibrium set of prices is very complicated and limited once one tries to include an assumption allowing increasing returns to scale – which must necessarily be allowed whenever imperfectly competitive firms are included in a general equilibrium model.

### 5.4.5. Operational?

Machlup was eventually criticized[20] for claiming that 'Equilibrium as a tool for theoretical analysis is not an operational concept; and attempts to develop operational counterparts to the construct have not been successful' [1958, p. 11]. By his saying the idea of an equilibrium is not operational, he meant that it is not empirical and possibly not refutable if false. But few critics who are members of the older culture ever complain about equilibrium models not

---

17. This was illustrated in Figure 3.3.

18. As Figure 3.3 shows, at the lowest average cost, average cost equals marginal cost and thus in the close neighbourhood of that point average cost is constant (i.e., horizontal). This also implies at that point that the returns to scale are constant (i.e., locally independent of the level of output) when profit is zero – this will eventually be the case whenever competition is perfect but not be so when competition is 'imperfect'.

19. Again, I will return to consider some further implications of such increasing returns to scale in Part II.

20. For example, see Finger [1971].

being operational. After all, there have been applied or computable general equilibrium (CGE) models for many years.[21] Of course, operationality was never an issue for the old culture given that Wassily Leontief was building empirical-based input-output models for years. However, it was an issue for those of the newer culture – those building and analyzing formal equilibrium models.[22] Samuelson said that his mathematical analysis would be concerned with what he called 'operationally meaningful' statements and propositions. These were merely code words for statements capable of being falsifiable 'if only under ideal conditions' [Samuelson 1947/65, p. 4]. Samuelson did not explain why this was necessary but it can be explained. Those who were on the front lines of promoting mathematical model building in the late 1930s were criticized by those claiming that mathematics could only produce tautologies.[23] Samuelson's PhD thesis was intended to prove those critics were simply wrong because if a statement can be proved to be conceivably falsifiable, it cannot be a tautology.[24] So again, while operationality of equilibrium models may be a concern for the formal mathematical model builders, it rarely is ever a concern of the critics of the realism of such formal models.

Having said this, I think the critics of what Machlup [1958] was saying did not understand what he was saying. He was not talking about equilibrium models – he was talking about the concept of equilibrium. And to a certain extent he is right. One cannot observe an equilibrium directly and that is why we build equilibrium *models*. It is only the observed mathematical status of the variables of the equilibrium model which defines whether or not there is a state of equilibrium.

21. CGE models are in one sense the modern version of input-output models from the mid-1930s. CGE models are empirical to the extent that they combine the system of equations with observational data about the included variables. These models are most often found in studies of agriculture and development economics. I will discuss these more in Chapter 10.

22. A popular early example of such activity was Samuelson's that was published in his PhD thesis – see Samuelson [1947/65].

23. For those unfamiliar with formal logic, a tautology is a statement for which one cannot conceive of a counter example – for example, the statement 'I am here or I am not here' is true regardless of who 'I am' or where 'here' is. Note that tautologies should not be confused with analytically true statements which are true by definition of the non-logical words. In this example, the logical words 'am', 'not' and 'or' are not considered matters of definition.

24. Economists too often say economic theory must be testable in order to avoid tautologies but this is asking too much. All that is necessary is that a theoretical proposition be *conceivably* false to assure that it is not a tautology. Performing a test involves much more in terms of specifying the requirements for a test of a proposition; see further Boland [2014, ch. 9].

## 5.5. EXOGENOUS VS. ENDOGENOUS VARIABLES IN EQUILIBRIUM MODELS: CAUSE VS. EFFECT?

There is another remaining characteristic of equilibrium models which needs to be discussed. It has to do with the two kinds of variables in every equilibrium model: exogenous and endogenous. As I discussed in the Prologue, they are distinguished in terms of what the model purports to explain. The model purports to explain the identified endogenous variables and does not explain the status or values of any of its identified exogenous variables. It is tempting to call the exogenous variables 'givens' but that can be misleading when you consider equilibrium models based on Marshall's distinction between short-run and long-run equilibria. I say it is misleading because as I explained in Chapter 1, some of the givens in the textbook short-run microeconomic model might be exogenous givens only in the model's short-run since they are included among the endogenous variables in the Marshallian long-run model. For example, productive capital (i.e., machines) may be a given and fixed variable by Marshall's definition of a short-run equilibrium but the quantity acquired by a producer is an endogenous variable to be explained in Marshall's long-run equilibrium.[25]

It is also tempting to consider the exogenous variables to be 'causes' and the endogenous variables to be 'effects' but this approach has problems whenever there is more than one exogenous variable in the model. Let me explain.

### 5.5.1. Causality among economists

The only time in the last five or six decades when causality was a burning issue in economics was in the 1970s and 1980s.[26] The issue then was whether inflation was caused by there being too much money in the economy. And as such, this was purely an ideological debate between followers of Milton Friedman and followers of John Maynard Keynes. I suspect that among economists, causality matters only when ideology matters. Economists interested in theory more than ideology will be more satisfied with the alternative: the commonly used distinction between endogenous and exogenous variables that I have been discussing. As I have been stressing, endogenous variables are the ones 'determined' by a conjectured equilibrium model and the exogenous ones are not. While there was a time in the 1950s when this distinction had its roots

---

25. And this includes explaining the size of the producer's industry since his long-run allows free entry and exit from industries.

26. The remainder of this chapter is based on Boland [2010], a review of Cartwright [2007].

in the ordinary notion of causality, today it does not. So, let me explain why economists see no need to be concerned about causality.

### 5.5.2. Causality and economic model builders

Boland [2014] was devoted to examining model building today and it discussed the two main types of model. One type of model is called *theoretical* simply because these models do not involve econometrics or empirical data although usually they otherwise involve mathematics.[27] Econometrics is the usual characteristic of the other type of model, which is called *empirical*. While it is common to distinguish between these two types of model, it is important to recognize that all equilibrium model builders respect the endogenous-exogenous distinction between types of variable recognized in their models. Such recognition is not on the basis of which type of variable 'causes' which type of variable, but on which variables are 'determined' (and hence explained) *by the model* and which are not.[28] There are also two other important distinctions playing a role: One is between elements of a model that are observable and those that are not; and the other is between elements that are fixed over time and those that are not.

For the easiest way to illustrate how model builders use the distinction between endogenous and exogenous variables one need only look at the macroeconomic equilibrium model found in the typical macro textbooks. For a very simple example textbooks often use the common and old Keynesian explanation for the aggregate level of the national income, $(Y)$, which today is called the Gross Domestic Product (GDP). The key behavioural equation is $C = \alpha + \beta \cdot Y$. The $C$ might represent the level of a year's aggregate consumption expenditure and the $Y$ might represent the level of that year's GDP. For such a simple model, more is needed, as usually the model builder would be trying to explain both consumption and income. So, another relationship (in this simple case, an identity that defines $Y$ specifically) would be added, such as: $Y = C + Z$ where $Z$ might represent the level of that year's aggregate investment (perhaps including private investments in production capital as well as government investments in such things as infrastructure). In this very simple model (consisting of only these two relationships), $C$, $Y$, and $Z$ represent observable variables. As a matter of simple algebra, one could 'solve' for the

---

27. For those unfamiliar with graduate economics, econometrics involves building a model and then using extra mathematical tools to statistically estimate the values of the parameters of a model using numerous empirical observations. Think of using such a model to explain why and how a nation's income has been changing over many years.

28. Of course, occasionally one can find economic model builders using the word 'cause' but this will be only as a synonym for mathematically 'determine' and meaning nothing more.

posited endogenous variables, consumption (C) and income (Y) if one knew the values of the parameters α and β as well as the one unexplained (hence exogenous) variable Z. A more complicated model might explain Z with a third equation and render it another endogenous variable but that model would then require another exogenous variable because without at least one exogenous variable, the model would be indeterminate or, worse, circular.

Note that in such a simple model, nothing would be said about which variables or parameters are fixed in time. And, it would not be necessary if one were only interested in the one set of observed variables (C, Y and Z) for one point in time (e.g., at the year's end). But usually, such a model would be used to explain the endogenous variables at many points in time (i.e., for many years). For this purpose, it is assumed that the parameters α and β are ergodic (that is, they are assumed not to change over time), each at its single value which will span the period of time being considered. When Keynes was explaining the level of aggregate consumption, he explicitly posited that the β represented a psychologically given 'marginal propensity to consume'.[29] But this propensity is not explained; it is just assumed to be represented by a fixed and unobservable parameter. And most importantly, it is a fixed parameter, not an observable exogenous variable.[30]

It is important to distinguish between theoretical and empirical models since economists who would build such a theoretical model are just assuming fixed known values for α and β – but econometricians who would build an empirical model will make no such assumption. Instead, the objective of building an econometric model is to determine – by means of a type of reverse engineering – the fixed values of the ergodic α and β using the observed values of the endogenous and exogenous variables over the time-span of interest. Since observations are rarely accurate, econometricians of course add 'error terms' to all of the equations representing the assumed relationships but usually assume that over the time-span in question the mean value of each error is zero. Many econometricians view their models – once the econometrically determined values of the parameters are substituted for α and β – as true 'generators' of the observed endogenous variables.[31] If the model is properly

29. Specifically, β says that for every extra dollar of income, each individual spends a fixed percentage – i.e., a fixed percentage of each dollar of income received during the whole span of time encompassed by the years considered.

30. Actually, the α and β are mathematical artifacts of the linearity assumption; Keynes was just suggesting what he thought was an obvious interpretation of the β. I will return to discussing Keynes' use of β in Chapter 14.

31. That is, if you were to plug the econometrically determined values of the parameters as well as the values of the observed exogenous variables into the model at each point in time, the deduced values of the endogenous variables would be the ones observed allowing only for the error term.

specified, only one set of values for the parameters will be found by means of the reverse engineering.[32]

Now note well: in this elementary exposition about economic models and econometrics there is no mention of 'causes' or 'causality' and as far as most equilibrium model builders are concerned, there is no need for it! Beside the avoidance of ideological stances, there are at least two other reasons for this. The main one is that most equilibrium models involve many exogenous variables and many parameters; hence realistically no one of them could be deemed 'a cause'.[33] The other is that today's formal equilibrium model builders are more interested in the mathematical properties of their models than in anything they might deem to be metaphysics. That is, mathematical relations such as $Y = f(X)$ says nothing about whether $X$ causes $Y$ or the reverse. Even if one wishes to say $X$ causes $Y$, the math by itself can never tell us that. It is just a way of interpreting the math. In a 1953 article, Herbert Simon tried to specify a special case where the mathematical structure of the model could be used to force a particular interpretation. But, only in a very limited sense is his structural approach about the idea of causality. As Kevin Hoover [2001, p. 37] explains:

> Simon [1953] offers a useful account of causal order in a system of simultaneous linear equations. One variable in such a system is said to cause another if one must know the value of the first variable in order to solve for the value of the second variable. Simon's notion of causal order is one of a hierarchy of nested subsystems in which the causes are located in the more central subsystems and the effects in the more peripheral subsystems.

Having to know the value of one variable in order to determine the value of another variable is merely an artifact of the mathematical matrix form Simon invents and nothing more.

And in Chapter 9 of his 2001 book, Hoover' goes on to explain that there is, however, one possible causality interpretation when it comes to a model's exogenous variables. For example, if an exogenous variable represents the level of government investment, one could speak of an investment multiplier. This, of course, implies a level of control over that exogenous variable. Equilibrium models designed to represents a type of hypothetical, intellectual experiment are useful for evaluating investment multipliers.[34] To see this, consider

---

32. That is, a properly specified model is said to be 'identified' – but I will not bother with this issue here other than to say that most econometricians think a properly specified model must have the same number of exogenous variables as there are endogenous variables.

33. I think it might be helpful for some purposes to just consider the exogenous variables to be 'influences' rather than 'causes'.

34. And as I explained in Chapter 1, the multiplier notion is basically a version of comparative statics analysis.

a model which explains national income (among other things) and has, as one of its exogenous variables, the level of government investments.[35] Such a model can be used to calculate how much national income would change if, *ceteris paribus*, the level of investment were to change. To do this one solves for the level of national income ($Y$) as a function of all the exogenous variables and, with an equation representing this solution, one performs the intellectual experiment[36] and thereby determines the potential effect of a change in the level of government investment. Note, however, it is all too easy to mathematically manipulate that equation and see it instead as a solution for the level of investment as a function of all the other variables. I note this only to underline that there is nothing in the mathematics about causes to prevent such a reversal even though the intention of the model is to be able to say that a change in investment could be seen to 'cause' a change in income. Of course, if the model is a true representation of the real economy in question and if one could control all of the exogenous variables (and not just a policy variable such as the level of government investment) so that *ceteris paribus* can be assured, then one might be able to make a causal claim. But most economists today would find this an unrealistic stretch although it usually is accepted for heuristic purposes in classrooms.

### 5.5.3. Can economists so easily avoid causality?

Contrary to how most economists avoid using any concept of causality in their equilibrium model building activities (again, they are only concerned with whether their model can be used to mathematically determine the values of endogenous variables), I think every theoretical equilibrium model can be seen to be positing a causal mapping from the vector space of values for the exogenous variables to a vector space of values for endogenous variables. The posited relationship between the parameters and all those variables is what constitutes the mapping (as I explained in Boland [1989, ch. 6]).[37] What I proffered is that every model posits a finite list (i.e., a vector) of exogenous variables for the purpose of explaining the values for the list of (possibly observed) endogenous variables. Implicitly every equilibrium model claims that the list of exogenous variables includes the only ones (through the posited mapping) to be determining the values of the list (i.e., vector) of endogenous variables.

35. The variable $Z$ in the simple model above would serve such a purpose.
36. This would be by performing a calculus differentiation to determine the change in $Y$ as result of a change in $Z$. And as I said, this could be considered merely an exercise in comparative statics that I discussed in Chapter 1.
37. However, note that in that chapter of that book I was mostly concerned only with the falsifiability of such a mapping.

The theoretical claim is that no other potential exogenous variables matter. If we think of a point in the vector space of exogenous variables as a finite vector of their values, then we could say that that *vector* plus the mapping are together the cause of the (observed) *vector* of the endogenous variables. Note also that this vector-based interpretation has nothing to do with probabilities.[38] If the values of the endogenous variables are to be explained, then the mapping and list of variables will constitute a conjecture both about the mapping's realism and the claim about which potential exogenous variables are the only ones that matter. As a conjectured explanation, every equilibrium model can be seen to be nothing more than such a mapping between vectors.

Of course, seeing an equilibrium model to be a mapping between vectors might appeal to the newer culture I have been discussing, but I doubt any member of the older culture would be found jumping on this bandwagon. The idea of causality is usually relevant for members of the older culture, for whom a state of equilibrium would always be about something we could see out our windows. Yet few in the newer culture would see a need to see their models to be about causality since, as I noted earlier, nothing in the mathematics can say anything about causality – causality, they will say is only someone's interpretation of the mathematics used in their models.

---

38. Probabilities are recognized in the econometric estimation process because observations used in the econometric estimation process are rarely accurate or exact, and in some cases are used to compensate for possible missing variables that might have a small effect on the econometric estimation of a model's parameters.

# PART TWO

# The limits of equilibrium models

# CHAPTER 6

॰⋏॰

# Recognizing knowledge and learning in equilibrium models

Recognizing knowledge and learning has been a concern for decades as the following quotations demonstrate. First there was Hayek [1945, pp. 524 and 526] who observed:

> If we can agree that the economic problem of society is mainly one of rapid ad-
> aptation to changes in the particular circumstances of time and place, it would
> seem to follow that the ultimate decisions must be left to the people who are
> familiar with these circumstances, who know directly of the relevant changes
> and of the resources immediately available to meet them .... Fundamentally,
> in a system where the knowledge of the relevant facts is dispersed among many
> people, prices can act to coordinate the separate actions of different people in
> the same way as subjective values help the individual to coordinate the parts of
> his plan.

About Hayek's view, Lachmann [1976, p. 55] noted:

> In modern Austrian economics ... we find the problem of knowledge to be a
> matter of fundamental concern. In 1946, in criticizing most modern theories
> of market forms, [Professor Hayek] pointed out that competition is a process,
> not a state of affairs, and that it reflects continuous changes in the pattern of
> knowledge.

And in the next decade, Lachmann [1982, p. 635] added:

> When Professor Hayek, ... in presenting 'Economics and knowledge', sug-
> gested that the most important task of economics as an empirical science

consists in explaining how men come to acquire knowledge of the 'data' governing the markets in which they operate, ... the whole problem was stated in equilibrium terms.

So, from early on, Hayek like most Austrian economists recognized the importance of individuals' knowledge when talking about market determined prices – as we see in these quotations (and as Lachmann explicitly observed). Hayek is often recognized for his insistence that if one is going to use equilibrium models to explain any social event and be faithful to methodological individualism,[1] one must explain within the model how the relevant individuals obtain the knowledge needed to assure that the equilibrium will be attained. How knowledge and learning is included in equilibrium models will be the subject of this chapter.

## 6.1. MODERN ATTEMPTS TO INCLUDE KNOWLEDGE AND LEARNING

The current interest in game theory analysis seems to offer a new opportunity to address the question of relationship between knowledge and so-called rational decision making with an obvious purpose. As Lawrence E. Blume and David Easley [1998, p. 63] observed:

> Game theory presents learning issues similar to the issues of expectation formation in economies with a sequence of incomplete markets and markets with differentially informed traders. In games with incomplete information, a (Bayes-Nash) equilibrium implies that, throughout the course of play, players will be learning.

Some game theorists are content to merely assume that knowledge and learning can be addressed in a Bayesian manner (e.g., a player starts with an apriori probability assessment and uses new observations to update the assessment). However, some other game theorists think such an expedient as Bayesian learning can be problematic.[2] The question for all the modern equilibrium models is whether the process of learning attributed to the agents in the model can be reconciled with the achievement of an equilibrium in a

---

1. As I explained in Boland [2003, ch. 2], this is the often recognized methodological requirement for every neoclassical economic explanation. It was first introduced by Schumpeter in 1909 who wished to recognize that social events are always the result of individuals making decisions and that this must be apparent in the explanation of those events. I will be discussing this requirement throughout this chapter.
2. See Bicchieri [1993], Mariotti [1995], or Albert [2001]. For an extended discussion of so-called Bayesian learning, see Boland [2003, ch. 8].

consistent manner.[3] Too often the achievement is assured only with excessive or otherwise unrealistic assumptions concerning the nature of knowledge possessed by the individual decision makers.

Game theorists can also be found embracing Herbert Simon's [1957] notion of bounded rationality, which is based on the claim that the acquisition of perfect knowledge would require learning abilities that no real human can have. But, if one is building an equilibrium model, bounded rationality with imperfect learning may also be problematic.[4]

The general notion of adaptive learning has also played a role in modern attempts to include learning in macroeconomic equilibrium models. There are many variations but all involve the notion that uses a form of forecast or prediction formula based on past observations and then when new information is obtained by using the formula, the formula is updated.

All of these various learning rules or devices are included in an equilibrium model to make it possible for the agents in the model to obtain the needed knowledge to be able to reach the model's equilibrium state. In most cases, the models are rather vague about what constitutes knowledge in these models other than that about the values of future prices or demands as in the case of rational expectations. Most models usually presume a quantity of information alone is the basis of knowledge. And most important, all presume a theory of learning that was refuted 200 years ago.

## 6.2  RECOGNIZING KNOWLEDGE
## IN EQUILIBRIUM MODELS

Long ago I began teaching my students that they should consider knowledge to be like health rather than like wealth. That is, knowledge is something that can be *improved* – it is not something quantifiable that you can have

---

3. I will postpone to Chapter 9 any extended discussion of the various ideas model builders employ to characterize learning in an equilibrium model.

4. For those readers unfamiliar with Simon's view, his bounded rationality is best understood as an application of his recommended 'satisficing' as an alternative to maximizing. Compared to maximizing utility, satisficing amounts to setting a minimum satisfactory level of utility which may happen to be less than the unknown maximum level. This leads to a range of acceptable choice options. The bounds on that range define bounded rationality. For more on this and its problems, see Boland [2003, ch. 4].

It should also be pointed out that there are macroeconomic model builders who claim to use bounded rationality but are not using Simon's version. They are instead using a version created by Thomas Sargent who attributes the boundedness to agents using assumptions that are not exactly true and so any predictions or deductions will also not be exact. For a discussion of Simon's versus Sargent's bounded rationality, see Sent [1997].

*more of.*[5] Having long been a student of Karl Popper's philosophy of science, this distinction seemed obvious. But the quantity-based view of knowledge and learning is so commonplace that it is difficult for most economic model builders to consider any alternative. Knowledge is not information. Knowledge consists of the theories we think are true and exist to explain and thereby understand potential observational information. So, why do so many people think knowledge is information and hence a quantity one can have more of?[6]

It might obviously be asked, how can we conceive of an alternative to the quantity-based theory of knowledge and learning? One alternative view of knowledge and learning can be traced back at least as far as the Socrates of Plato's early dialogues.[7] At root, this Socratic view of knowledge and learning considers one's knowledge to be like one's health and not like one's wealth. Specifically, one learns by improving one's knowledge (viz., by eliminating one's erroneous knowledge) rather than by accumulating more knowledge (e.g., finding more confirming evidence in favour of one's knowledge[8]). Again, the main problem with the typical equilibrium model builders' reference to knowledge is then that all too often they confuse knowledge with information.[9]

---

5. Of course, one can have knowledge of more things but that is not what equilibrium model builders usually mean by having 'more knowledge'. And, one can think of a way 'more knowledge' makes sense – specifically, more knowledge of a language or knowledge about more things or more topics such as knowledge about physics and chemistry and geography, etc. But, this is a matter of scope, not a matter of quantitatively more knowledge about a single topic or subject.

6. For an interesting discussion of how a confusion between alternative views of knowledge can cause many theoretical problems in economic models, see Binmore [2011].

7. The best example would be Plato's dialogue 'Euthyphro' but the 'Apology' and 'Crito' are also worthy of consideration.

8. Since, even with confirming observations, this knowledge may eventually be found to be false.

9. This confusion between knowledge and learning has a long history and has to do with false beliefs in inductive learning as I briefly noted in note 22 of Chapter 1. The quantity-based view of knowledge and learning has been around for 350 years and remains despite its being refuted more than 200 years ago. Specifically, people take for granted the induction-based theory of knowledge and learning – a theory philosophers attribute to the seventeenth-century philosopher Francis Bacon and its refutation to the eighteenth-century philosopher and economist David Hume. The imaginary induction-based theory of knowledge presumes one's knowledge is nothing more than the observations (hence information) that one can use to *deduce* the truth status of one's knowledge – supposedly, this knowledge has been induced by the collection of observations. Unfortunately, there is no such logic that allows for such an induction or deduction – that is, for a deduction using only observations and not using any general statements (like *all* producers are profit maximizers).

According to Bacon's old theory of knowledge that was presumably based only on induction, one *acquires* knowledge by making observations or, more generally, one's knowledge is accumulated experience. In other words, knowledge is nothing more than a summary of *past* observations or experience. Some philosophers call this

In one sense we cannot avoid a quality vs. quantity distinction here. Clearly, whenever knowledge is seen to be just information, it is viewed as a quantity and thus it can be amenable to quantitative treatment and analysis. For example, consider again Stigler [1961], the famous article 'The economics of information' that, among other things, can be seen to be about one's knowledge of the best available price for a desired product. As I mentioned in Chapter 1, his article portrays learning to be accumulating observations that allow better and better probability-based estimates of the parameters of the distribution of a market price of interest. The greater the quantity of observations, the smaller will be the standard deviation of its estimated mean. Quantifying knowledge this way makes it easy to conclude that observational-information-based knowledge is an economics issue. Namely, if one has to pay for the information (i.e., for the observations), then better estimates will be costly such that an optimum is reached when the marginal improvements in the estimate will not be sufficient to justify the marginal cost of the next observation.[10] At best, Stigler's views of knowledge and the role of information are limited to just one type of knowledge, namely, knowledge of the true value of one variable such as the best price available. Assumptions about how one learns about the true nature of one's preference map[11] would seem to be far more complicated. Nevertheless, rejecting the quantity-based theory does not necessitate rejecting Stigler's analysis, per se. But, it does beg the question about what we assume concerning how a decision maker interprets the quantity of information.[12]

quantity-based view of knowledge and learning the 'bucket theory of knowledge' on the grounds that the more observations you make, the more knowledge you have. According to this quantity-based bucket theory of knowledge, not only is one's knowledge merely the contents of one's bucket, but one learns only by adding more to the contents of one's bucket. The problem of relying on quantity-based knowledge is the subject of Boland [2003, ch. 1]. While that discussion is more concerned with understanding the methodological aspects, here the discussion will be concerned more with the theoretical implications of that discussion – particularly when it comes to assumptions about learning and knowledge installed in equilibrium models.

10. In Stigler's day that might be in terms of coins placed in a pay telephone. Today we might think of it as the cost of extra minutes for one's cell phone. In effect, Stigler was responding to critics of equilibrium models who claimed the state of equilibrium required perfect knowledge. And Stigler's response was simply that perfect knowledge would be too expensive and thus non-optimal!

11. For those unfamiliar with this jargon, when explaining a consumer's utility maximizing choice, economists use a map to represent how the consumer's preferences change depending on how much has been consumed so far. The map can also indicate the utility (or satisfaction) obtained when consuming any bundle of goods at a point on the map. Seen this way, the map is seen to represent what is called the consumer's utility function. More will be explained about all these later in this chapter.

12. For an even more detailed discussion of how a decision maker might be assumed to interpret the relationship between observations and knowledge claims, see Boland [1986, ch. 11; 1992, ch. 11; 2003, chs. 1, 12; and 2014, ch. 11].

## 6.3 TOWARDS INCLUDING REALISTIC LEARNING IN ECONOMIC EQUILIBRIUM MODELS

In his 1959 article, Richardson urged [p. 229]:

> If we assume that consumers are fully informed about their own preferences and about the properties of all the goods they may buy, then the models we construct will indeed suffer in terms of realism and will offer a less than perfect account of the workings of the actual economy.

It may be too early to say, but I suspect that progress in including a realistic notion of knowledge or learning will not be easily obtained. And whether it is with game theory analysis that uses bounded rationality or with obvious quantity-based notions of knowledge such as that presumed by the use of adaptive learning or Bayesian learning, this will be so until these notions are abandoned in favour of a Socratic improvement-based conception of knowledge and learning.[13] I discussed these conceptions of learning and knowledge in the previous section where the point was that quantity-based learning treats knowledge like wealth – something that one can have more of – whereas improvement-based learning treats knowledge like health – something that one can improve by first recognizing one's errors.

Whether one's conception of knowledge matters depends directly on the equilibrium model one is trying to build. Equilibrium models are particularly troubling when it comes to matters of knowledge. How or whether one's conception of knowledge matters will be examined with three diagnostic questions that can be used to determine when the issues discussed here can matter. For each question, I will try to indicate what I think should have been learned from this discussion.

### 6.3.1. Does learning matter in the model?

If one is going to build an equilibrium model such that learning matters, I of course think one must reject the quantity-based theories of knowledge and learning in favour of the Socratic improvable-quality-based view of knowledge and learning (whereby one learns by discovering one's errors). That is, one must reject any theory that equates learning with just the accumulation of data and instead must adopt the view that learning is error correction. The first step in adopting the Socratic view is simply to recognize that every individual decision maker holds one or more conjectural theories about the

13. For more about the Socratic view of knowledge and learning, see Boland [1997, ch. 20].

various elements of the decision situation faced. Moreover, and most important, these theories are possibly false.[14]

That knowledge can be seen to be manifested in theories is not as demanding as might at first seem. The notion that decision making is a *process* rather than an instantaneous *event* should be obvious and thus, as Richardson argued in his 1959 article discussed in Chapter 4, every decision that in any way depends on a future state of affairs involves conjectural expectations formation.[15] At a minimum, what the Socratic view does is to extend the notion of theoretical conjectures to all of the knowledge requirements of decision making.

Extending the notion of theoretical conjectures to all knowledge requirements means that, for example, the consumer can be assumed to not know a priori what the true shape of his or her preferences map is. Instead the consumer can be assumed to conjecture what he or she would expect to be his or her reaction to consuming a particular bundle (of commodities) or switching from one particular bundle to another. When assuming that a consumer has a particular type of preference map, an equilibrium model builder is in effect merely assuming that the consumer conjectures that type of preference map and that the conjectured map will turn out to match what would be confirmed if he or she had the time to try out all of the infinity of possible or at least conceivable bundles.[16] The quantity-based theory sees the consumer to be accumulating data with each purchased bundle to confirm the a priori conjecture but this view of learning is incapable of dealing with refuting data. The Socratic view explicitly considers refuting data as potential learning opportunities, so how the decision maker deals with refuting data has to be made an essential part of the explanation of the decision maker's behaviour that is being modelled – just as Clower did in his 1959 article about the ignorant monopolist whose false conjectures led to expectation errors and adjustments to subsequent conjectures. That is, as Clower showed, (Socratic) learning matters if the process of reaching an equilibrium is to be explained.

### 6.3.2. What role do probabilities play in the model's decision maker's learning process?

So far I have not said much about the common notions of 'uncertainty', 'risk', or 'probabilities' beyond what I discussed in Chapter 5. Here I wish to point

---

14. And as I demonstrated in Chapter 3 with Clower's ignorant monopolist's possibly false conjectural theory about the shape and location of the demand curve it faced.

15. Again, his 'rational expectations' are nothing more than theoretical conjectures – 'rational' only in sense that obvious false conjectures are excluded.

16. For more about a consumer's preference maps or utility functions as a consumer's conjecture or assumption rather than as a psychological given, see Boland [2003, ch. 8, p. 138].

out that the introduction of any of these common notions is a direct consequence of the equilibrium model builder's attempt to avoid giving up the quantity-based theory of knowledge and yet still recognize the fallibility of a decision maker's knowledge or expectations. Again, the issue missing in any equilibrium model based on the quantity-based theory of knowledge and learning is that there is no way to deal with refuting data. When equilibrium model builders resort to probability-based notions of fallible knowledge, the model builders are inadvertently making explicit recognition of learning virtually impossible. Whenever the model's agents do learn by discovering and correcting their errors and the model builder assumes they use probability notions, it makes learning arbitrary or at least makes it very difficult for the model's agents to know when they have made an error.[17]

The problem here is not with probability notions but with the quantity-based theory of knowledge and learning. It is important to note that the Socratic view of knowledge and learning does not preclude the use of probability notions. It nevertheless does require that the equilibrium model builder be explicit about how their model's agents incorporate probability notions in their decision making process. Consider for example what kind of evidence would cause the agents to determine that their knowledge of the situation is false. Also, what theory do the agents hold concerning data handling? Do the agents think one can answer non-stochastic questions with statistical analysis?[18]

### 6.3.3. Does the equilibrium model involve agents' making decision errors?

One must recognize the necessity of addressing the possibility of decision errors once one recognizes the fallibility of all knowledge – particularly knowledge that is necessary in the process of decision making that might lead to an equilibrium state. To address the possibility of a model's agent making errors the model builder must first deal with how the agent becomes aware of an error and then deal with how the agent responds. This task is made much easier if it is recognized that at least some agents are aware of the fallibility of their knowledge and thus they treat any decision as a test of their knowledge. For example, in such a case a consumer is never certain that the choice made

17. The inherent methodological difficulties with testing probability-based models are examined in Boland [2014, chs. 9–10]. Also, various ways to model how decision makers learn are examined in Boland [1992, ch. 11], which addresses how decision makers respond to expectation errors or failed knowledge used to form the expectations.
18. Here stochastic merely means random or statistical variation is involved and non-stochastic means neither is involved.

is the one which will maximize utility. The strategy employed by the consumer will depend on the theories held by the consumer. In the simplest case, following what is found in microeconomic theory textbooks about the consumers choice situation when choosing a bundle of quantities of goods (X and Y) as illustrated in Figure 6.1,[19] the consumer may assume his or her indifference or preference map is 'strictly convex' to the origin[20] and that his or her choice has no effect on the price. In this case, the consumer merely searches along the budget line[21] first by trying out two widely spaced points (such as points B and C) and then testing his or her theory by buying a point between them. If the consumer's theory is correct, the third point will usually be better than the first two. In such a sequence of trial and error, the consumer can narrow down the choice to the one (i.e., point A) which according to the convexity assumption would be the utility-maximizing bundle. However, if the consumer's theory is false (either the map is not strictly convex or prices are not fixed),

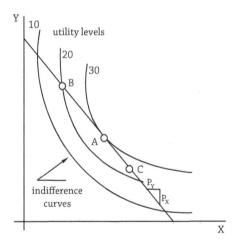

**Figure 6.1.** A consumer's preference map

19. For the benefit of readers unfamiliar with what is found in microeconomic textbooks, I have added more information than what is usually found in the usual textbooks' version of Figure 6.1.

20. For those unfamiliar with this jargon, it refers to a mathematical property of the utility function or preference map. If one uses a preference map (such as shown in Figure 6.1 which looks like a topographical map showing various altitudes) to represent the utility function and then draws a straight line between any two points on a single indifference line (which, like a line on a topographical map, is a line for which all points on the line yield the same level of utility) then all points on the line between those two points will be preferred to those two end points.

21. For those unfamiliar with this jargon, a budget line represents all the points or bundles of goods that the consumer could buy by spending his or her whole budget. Being a straight line reflects the assumption that the consumer's choices do not affect the going prices.

the consumer may not easily be able to narrow the choice to the true maximizing point or, worse, choose what turns out to be a non-maximizing point on the map.[22] In the case of such a failure to maximize, if the consumer is aware that an error has been made, the consumer would have to determine the source of the error.[23] Such a determination or even the error awareness seems to be beyond the textbook theory of the consumer. Obviously, whenever the consumer thinks his or her behaviour has an effect on the given prices, a much more complicated decision strategy would have to be involved.

To make sure what I am discussing here is clearly understood, let me present this issue of Socratic error awareness in a diagram not found in textbooks – specifically, Figure 6.2.[24] This diagram is intended to show what I am talking about concerning consumers having to conjecture their preference maps because they do not know a priori their true preference maps. It also can illustrate that as a result having to conjecture their maps, each consumption decision they make is a test of their conjectural knowledge as I suggested above. So, Figure 6.2 illustrates how a consumer's conjectured map that is found in all microeconomics textbooks can deviate in both shape and location from the consumer's true indifference map.

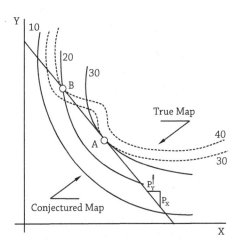

**Figure 6.2.** A true preference map vs. a conjectured map

22. Much like Clower's ignorant monopolist ended up by choosing a supply quantity that cleared the market but was not truly the profit-maximizing supply even though it was thought to be since marginal cost equaled the falsely conjectured marginal revenue.

23. A good place for the consumer to start, of course, would be examining the conjectures about the decision situation – in the case of the consumer, the conjectured preference map.

24. This diagram is similar to the one I used in Boland [2003, ch. 8].

So, in Figure 6.2, the consumer is *assuming* his or her utility function is the textbooks' strictly convex map represented by the solid lines when in fact the true indifference map is the one represented by the dotted lines that is not strictly convex.[25] If the consumer starts his or her trial and error approach in the close neighbourhood of point A, the consumer can thus end up erroneously choosing point A even though point B is where utility would actually be maximized.[26] Error awareness is about how such a consumer *in an equilibrium model* would learn that he or she not maximizing.[27] This is particularly problematic whenever the model follows the standard textbooks by assuming all consumers' preference maps are psychologically and hence exogenously given and thus beyond question.[28]

Surely, modelling error awareness can be easy or difficult. The easy case is when the consumer in a trial and error sequence finds that the third point is not preferred to the first two. Similarly, Clower's monopolist going to the market expecting one price level and finding the price turning out to be different involves a direct awareness of an expectation error. As I discussed in Chapter 3, knowing what is the source of the error is a more difficult question. The behaviour of Clower's ignorant monopolist perfectly represents the problem of error awareness. In Clower's Model III, if the market is cleared for the expected prices, there is no additional information available to indicate that the assumptions made by the decision maker are false and thus that maximization is not actually being achieved. When a decision maker must make assumptions prior to participating in a market, the question of error awareness becomes rather troublesome. And again, the textbooks' (viz., the quantity-based) theory of knowledge or learning is not very helpful. And possibly worse, even Hayek's [1945] apparent presumption that the market provides information that assures that all market participants learn what is necessary to guarantee the attainment of that equilibrium state is at best open to question.

---

25. The solid-line map is the one that textbooks and most equilibrium model builders consider to be psychologically given and hence the true map – but here I am saying the true map is unknown thus necessitating the textbooks' map to be seen as a conjectured map.

26. By considering Figure 6.2, one could ask the methodological question: Why do textbooks assume all consumers face strictly convex preferences (as shown with the solid lines)? One could easily answer this question: it is because such an assumption conveniently makes the textbooks' one behavioural assumption easily work to explain the consumer's choice.

27. This is necessary for the process of reaching the equilibrium. When the consumer becomes aware that utility is not being maximized, he or she will make a different choice. Moreover, not maximizing is not compatible with a state of equilibrium as any sub-optimality would necessitate any non-optimizing decision makers to change their decision and thereby upset the equilibrium state.

28. I will discuss the standard textbook's assumptions about preference maps more in Chapter 8 and in Part III.

## 6.4. THE PROBLEM OF MAINTAINING METHODOLOGICAL INDIVIDUALISM IN EQUILIBRIUM MODELS

Arrow began in his 1994 Richard T. Ely Lecture as follows [p. 1]:

> In the usual versions of economic theory, each individual makes decisions . . .
> In one way or another, these decisions interact to produce an outcome which
> determines the workings of the economy, the allocation of resources in short.
> It seems commonly to be assumed that the individual decisions then form a
> complete set of explanatory variables. A name is even given to this point of view,
> that of *methodological individualism*, that it is necessary to base all accounts of
> economic interaction on individual behavior. . . . The meaning of individualism
> was taken up . . . by . . . Joseph Schumpeter (1909), in a paper on the concept
> of social value.

What Arrow is referring to is Schumpeter's following discussion [1909,
pp. 216–17 and p. 231]:

> The only wants which for the purpose of economic theory should be called
> strictly social are *those which are consciously asserted by the whole community* . . .
> Many writers call production, distribution, and exchange social processes,
> meaning thereby that nobody can perform them – at least the two last named –
> by himself. In this sense, prices are obviously social phenomena.
> [I]t is here claimed that the term 'methodological individualism' describes a
> mode of scientific procedure which naturally leads to no misconception of eco-
> nomic phenomena . . .

For the most part, the recognition of knowledge or information in equilibrium
models is ultimately concerned with the knowledge possessed by individual
decision makers engaged in the process of determining what to buy, sell, or
produce. But textbooks usually only talk about the *market* determining prices.
So, does this mean that the knowledge of individual decision makers does
not matter? In this regard, while it might be tempting to see the textbooks
viewing the market as a 'thing' determining the price, if this were their view
it would violate that ubiquitous requirement of all neoclassical economic explanations – the one that Arrow is explaining in the above quotation and (as
noted in this chapter's first note) the one Schumpeter [1909] called 'methodological individualism'. I usually characterize this perspective on acceptable
explanations in neoclassical economics simply as 'things do not decide; only
individuals do'.[29] This requirement means recognizing that *any specific* price

---

29. For an extensive discussion of methodological individualism, see Boland [1982,
ch. 2] or Boland [2003, ch. 2].

marked on a price tag must be decided by someone,[30] but this recognition alone gives us no reason to expect *any particular* price to be placed on the tag.

As I explained in the Prologue, some attributes of the textbooks' concept of a market equilibrium obviously make it interesting for neoclassical economics. This is particularly true when one recognizes an autonomy of individuals that involves an implicit question of central concern in the context of what Arrow and Schumpeter call methodological individualism. A central virtue of using markets to explain prices is that a market equilibrium seems to allow for any individual's complete autonomy at the same time allowing an explanation involving the economy as a whole. However, it remains to be seen whether an equilibrium explanation of prices can always be constructed such that both complete autonomy is preserved and a logically adequate determination of market prices can thereby be made.

The textbooks' neoclassical equilibrium model does obviously maintain an important autonomy of individual decision makers while explaining the equilibrium of a market. At a minimum the textbook's neoclassical theory separates the determination of demand from the determination of supply. Apparently by separating demand from supply a minimum, but essential, element of autonomy for decision-making individuals can be built in by assuring the absence of collusion.[31] Moreover, for any particular set of prices being charged, the autonomous individual agent acts *freely* in deciding what, or how much, to demand or supply given those prices. Note well though, maintaining the separation of demand and supply is a decision made by the theorist – i.e., deliberately to facilitate methodological individualism. Much of traditional textbooks' theory has been developed to justify this separation and at the same time assure that the equilibrium reached in the long run will amount to Adam Smith's 'best of all possible worlds'. Of course, as Richardson suggested, it certainly would not be the 'best' should individuals encourage collusion or be dependent on each other's approval.

To better understand the textbooks' economic concept of a market equilibrium, I want now to consider this concept in a different way. As Schumpeter noted, our equilibrium theory of prices says that *prices are social institutions*. To say this, however, brings up in a new form the dilemma between assuring individuals' complete autonomy and providing logically adequate explanations of a whole economy. Historically, there have been two basic

30. Perhaps today we would instead say that some decisions are made indirectly by someone using a computer program but of course, some other individual would have had to program the computer to perform the decision.

31. In economics what I am calling 'autonomy' would usually be seen to involve the question of whether there is so-called 'consumer sovereignty'. For the general matters being discussed here, the autonomy of decision making is the main concern for all participants in the market – and for proponents of perfect competition this also includes the exclusion of any form of collusion since that would preclude autonomy in the usual sense.

views of social institutions and these two views are diametrically opposed. First, there is the one I have been talking about – the strict methodological-individualist view, which says that all institutions are merely aggregate manifestations of autonomous individual behaviour and hence institutions are explained *only* in terms of the behaviour of each and every individual.[32] For example, if prices are social institutions, then prices will be the equilibrium prices only if *everyone* agrees that the going prices should not be changed. Second, there is what has long been considered the main alternative – the strict holist view which says that some social institutions have an existence (and hence a determination) beyond the individuals that use or help create them.[33] According to the methodological holist view, individuals play no determining role but instead they just need to conform. In economics the holist view could be applied to prices. The common example would say that the real price will always be determined ultimately by its 'natural value' or its 'just value' or its 'labor value', etc. As such, individuals play no role in its ultimate determination. It is strict holism that is specifically rejected when textbooks reject 'natural' causes (such as labor embodiment) as *sole* determinants of prices.

Virtually all members of the old culture discussed in Chapter 5 seem to agree that the analysis of a state of equilibrium alone will never be sufficient to explain prices in a manner consistent with methodological individualism. As I discussed in Chapter 2, Arrow's 1959 article explained that what is needed instead is a clear explanation of the disequilibrium state in which the process of reaching an equilibrium state takes place.[34] Expanding our view of prices to include disequilibrium states as well as equilibrium states allows for individualism (the price-tag marker) and at the same time recognizes prices as holistic and endogenous short-run givens which constrain the individual's actions (e.g., by determining opportunity costs[35]). The individual sellers can pursue what they think is in their own interest but in the long run (a run long enough for general equilibrium to be obtained), they will find it in their own best interest either to all charge the going equilibrium price or to demand or supply the quantities that are consistent with the equilibrium price.

32. Economists often think that this must ultimately involve the psychology of the individuals; see Boland [1992, ch. 8]. I will return to this unnecessary view later.

33. The classic Marxist example of such an alternative is to say institutions may exist only for reasons of 'class interest'.

34. And this was also Paul Samuelson's [1947/65] message in his famous PhD thesis, *Foundations of Economic Analysis.*

35. For readers unfamiliar with this jargon, opportunity costs are the indirect result of using one's budget or available funds for a product purchased or an investment engaged in since doing so incurs giving up the opportunity to buy or invest in

## 6.5. THE PROBLEM OF THE COMPATIBILITY
## OF GENERAL EQUILIBRIUM AND PSYCHOLOGISM

What is needed now is to look closer at the question of what is meant by individualism, or the role of the individual, when talking about an entire economy being in general equilibrium. Again, as I discussed in the Prologue, all explanations based only on equilibrium models involve endogenous variables which the model purports to explain and exogenous 'givens' that influence the determination of those endogenous variables. A question not usually considered concerns how model builders determine what are acceptable 'givens' in their models of the consumer or the producer. To examine this determination, I will review the basic notions found in the typical Economics 101 textbooks. For most purposes, all that needs to be discussed is what is considered an acceptable equilibrium-based explanation in economics and has been for well over a century.

As explained in Chapter 5, neoclassical equilibrium models specifically identify two types of 'givens'. In the usual model there are: (a) those endogenous social variables such as prices that for the price-taking individual decision maker are givens (e.g., going prices, income distributions, wealth and capital distribution, wage-rates, etc.)[36] and (b) those exogenous variables that are supposedly 'natural' givens (e.g., tastes, availability of resources, learning abilities, biological growth rates,[37] etc.). In Marshall's neoclassical definition of the short run, individuals are unable to change any of these exogenous givens.[38] However, beyond the short run and into the long run, individuals can influence the long-run endogenous social variables, the (a) givens. The solution to the 'holist vs. individualist' dilemma discussed in Section 6.4 apparently lies here. In the short run, from the perspective of individual decision makers, prices are in effect holistic givens – as Schumpeter and Arrow indicate in the quotations at the top of Section 6.4. In the long run, however, more of the endogenous social variables become social consequences of individual

something else – for example, giving up the interest income one could have obtained by lending the spent or invested funds to someone else.

36. As I explained in Chapter 5, one needs to be careful about the word 'givens' since a variable may be an unchanging given for a decision being made by an individual but may still be a variable that is being explained for the whole economy – for example, the market prices I am discussing in the text. Another example is the amount of physical capital (e.g., machines), which is a fixed given in the short-run model but is an explained variable in the long-run model.

37. Although, even some of these are subject to research and developments directed at making these rates endogenous.

38. Exogenous variables for sure, but, again, in the short run, even some of the market's endogenous variable are fixed 'givens': for the individual, particularly the going prices that individuals face in the market; and for firms, of course, also the fixed amount of productive capital owned.

choices. In the textbook long-run equilibrium, the only exogenous givens appear to be the 'natural causes' or 'forces', particularly the presumed psychologically given tastes or preferences. For this reason, the type of methodological individualism in a neoclassical equilibrium is what some philosophers call 'psychologistic individualism',[39] since these psychologically given exogenous tastes and preferences are the only basis for the individuals' motivations in determining the demand decisions they make to maximize utility.

Consider again the simple neoclassical methodological-individualist equilibrium model considered in the Prologue. In this model's simple world whenever the equilibrium is unique, the explained set of values are said to be the only set which corresponds to the one particular set of values (or states) for the following exogenous 'givens': 'tastes' (which are represented by a psychologically-given preference map for each of the individual consumers), 'technology' (which following Marshall's mode of explanation is represented by fixed technologically-given production functions relating the individual outputs $X$ or $Y$ to the levels of inputs $L$ or $K$) and available resources (the Nature-given total and limited amounts of productive resources that exist in the world) – such as the total amount of labour or raw resources used to produce capital.[40]

If the long-run equilibrium has been reached *and* the (exogenous) 'givens' do not change, the long-run equilibrium values of the determined endogenous variables will *never* change! In other words, so long as the exogenous 'givens' do not change, our analysis is essentially static even though individuals may be thought of doing things continuously – such as changing inputs into outputs. Every week, each individual buys or sells exactly the same quantity of each good in its market because in this world there is no change in the endogenous demands or supplies without a change in at least one exogenous variable. Clearly then, in such a simple equilibrium model it would seem any interesting 'dynamic' analysis must somehow deal with changes in the exogenous 'givens'.[41]

Almost all economists credit Walras for attempting to specify the behavioural assumptions of a mathematical model of the whole economy that would logically and formally ensure the existence of a set of prices consistent with a general equilibrium of price-takers for any set of exogenous givens. Most equilibrium model builders who invoke what they call Walrasian general equilibrium theory are interested in a state of equilibrium where each individual

39. See Agassi [1960, 1975] – as well as Boland [2003, ch. 2].

40. Sometimes there is an additional natural given in the form of an 'interest rate', $i$, which may represent the opportunity costs of consuming today rather than investing in capital or other inputs to produce something for tomorrow (e.g., it may represent the Nature-given biological growth rate which follows planting of seeds).

41. Although, if we allow for learning in our model, this would be otherwise – as will be discussed in Chapters 12 and 15.

is autonomously maximizing subject to his or her personal constraints while facing the same set of prices as everyone else. In effect, for such an equilibrium model the determination of any set of equilibrium prices amounts to solving a set of simultaneous equations where the equations represent the maximizing conditions for each individual decision maker. As Wald [1936/51, pp. 369–70] pointed out, early Walrasian general equilibrium model builders often presumed that it is enough just to ensure that the number of equations equalled the number of endogenous variables recognized within the model.[42] But, the question is much more complicated.[43] If for no other reason, any real economy usually has an extremely large number of individuals and so the system of equations would be difficult to solve except in very special cases.

### 6.5.1. Multiple equilibria?

Even in the simple two-person model of the simple economy presented in the Prologue there are problems for the methodological-individualist interpretation of the neoclassical explanation of prices. No matter what decisions individuals in the model make in the process of reaching an equilibrium, for that model to constitute an explanation, there would have to be only one set of determined (endogenous) values for the given set of exogenous variables. As was explained in Chapter 5, for such a model this is the goal of a uniqueness proof. However, as explained in that chapter, if there is possibly more than one set of equilibrium values (i.e., more than one solution for the system of equations that comprise the equilibrium model), with that model we will not have explained why one equilibrium state would be reached rather than any other *logically possible* equilibrium state. However, if only one set of equilibrium prices is logically possible, does this mean that the givens are together the 'cause' of the market determined values and thus that our explanation of prices denies completely autonomous decision making to individuals? Unfortunately, it is difficult to see how the answer is not affirmative whenever the givens are considered fixed and thus unalterable by any individual involved.[44] If any

---

42. I have for years said this was Walras' view but that view is more a reflection of how model builders of my generation viewed Walras – without ever reading his work, of course. More importantly, as I noted in the Prologue, what is usually called the Walrasian general equilibrium model is really due to a simplified version created by Cassel [1918/23].

43. See Boland [2014, ch. 2].

44. Note that here I am talking about all the exogenous givens taken together, not individual causes. In Chapter 5, this was a case of talking about a singular vector of values for all exogenous variables causing another singular vector of values for all the endogenous variables. In this case a single vector could be seen to be causing a single vector.

'exogenous' variable is alterable by any individual recognized in the model, then it becomes an endogenous variable that needs to be explained. Doing so would necessitate recognition of one or more new exogenous variables in the model and as an explanation such alterability would lead to an infinite regress.

Clearly when there is just one set or vector of equilibrium values for the endogenous variables there is a serious problem for the usual adherence to methodological individualism. It would seem that if we know the values of all of the exogenous variables in the equilibrium model, there does not appear to be room for choice as only one set of endogenous variables can be realized no matter what we presume individuals in the model think they are doing. Can this philosophical obstacle be avoided or dismissed? Most economic theorists seem to think so either because at least they think they can ignore it or because they think there is an alternative. For example, some equilibrium model builders[45] simply accept as an alternative 'multiple equilibria', that is, accept more than one set of values for the endogenous variables which are consistent with the one fixed set of exogenous givens. The latter unfortunately is a defeatist position – no matter how broad-minded it may appear to be. Any hope of *explaining* the variables in question in terms of individual choices alone is conceded. But worse, if it is argued that there can be many possible sets of equilibrium values in any equilibrium model, then in effect every one of an individual's choices is *arbitrary* – arbitrary as a consequence of an arbitrary prior choice concerning which prices to take as his or her givens for making decisions for demand or supply. For some of us, such arbitrariness is just as bad as a denial of complete autonomy of decision making.[46]

This individualist dilemma can be addressed in another way by admitting that the 'givens' are not really given, since each can be influenced by individuals in the economy. Unfortunately, if carried too far – that is, if all the givens are made endogenous *within* our model of the economy, then the explanation of all variables becomes logically circular. One way to avoid circularity is to explain the 'givens' *outside* of the model in question. This approach, similar to that suggested by Thorstein Veblen at the beginning of the twentieth century, has been for the most part avoided except by a few economists who call themselves 'institutionalists' and who seem willing to take some institutions as exogenously determined but not by Nature-given exogenous variables.[47] Neoclassical economists for a long time rejected institutionalism, if for no

45. For example, Samuelson [1947/65, p. 49[ and Stiglitz [1975]; see also Kreps [1990, p. 108].
46. Multiple equilibria are also problematic if one is explaining observed prices. Any singular set of observed equilibrium prices means that all individuals must be making decisions consistent with that one set and not with any other of the possible multiple equilibria. Formal model builders might not be bothered by this, but the older culture discussed is Chapter 5 surely would be.
47. For further discussion, see Boland [1992, pp. 51–2].

other reason, because taking institutions as unexplained givens would undermine the methodological individualism of neoclassical theory by allowing
elites, power groups, government controls and other such holistic variables
to influence problematically the ultimate long-run equilibrium state.[48] Such
holistic influence might mean that the long-run equilibrium is not necessarily the 'best of all possible worlds' since it may only be the best for those with
holistic influence on social institutions. For now, I will leave these somewhat
philosophical problems aside and just recognize that methodological individualism is never rejected in mainstream neoclassical equilibrium models regardless of how troublesome some might find such considerations when it comes
to building equilibrium models that are consistent with the requirements of
methodological individualism.

### 6.5.2. Psychologism

The most commonly accepted approach to allowing certain givens to be
explained outside the model is to confess that since 'we are all humans', everything reduces to psychology. The requirement or expectation that all explanations of social events must be ultimately based on the psychology of the
individuals involved is the perspective I and others have called 'psychologism'.[49] This psychologistic perspective seems to have been the explicit view
of some economics writers such as John Stuart Mill in the nineteenth century
and maybe even the view of Vilfredo Pareto in the early twentieth century.
Today, it is at least implicit in most equilibrium economic models and certainly in most, if not all, elementary textbooks. In particular, it is often held
that strict methodological individualism would require us to explain even the
impersonal givens such as technology, resource availability, interest rates, or
wealth distributions, within any neoclassical equilibrium model. However,
given any variability of tastes or of other exogenous variable, some or all of
such nature or variability of individual tastes would have to be explained outside the model to preserve a minimum degree of exogeneity and avoid circularity. This 'psychologistic' method of allowing economists to explain everything except the natural givens goes virtually unchallenged in economics

48. For this reason, today we have 'New Institutional Economics' thanks to the
efforts of Douglas North [1978] and Oliver Williamson [1985] to address this problem. New Institutional Economics is the neoclassical economists' attempt to deal with
the observed fact that institutions exist and matter. Supposedly, these institutionalists would have us treat institutions as social entities influenced or maybe even determined by direct maximizing decisions made by individuals acting together over a
longer period of time rather than as exogenous institutions as in the usual neoclassical
model found in textbooks.
49. See Jarvie [1972] and Agassi [1975] as well as Boland [2003, ch. 2].

textbooks and literature since it still seems to be the only way to accommodate the demands of methodological individualism. But if not carefully considered, it is easily possible that relying on a psychologistic version of methodological individualism undermines what is meant by completely autonomous decision making[50] and in a long-run equilibrium or general equilibrium model completely undermines any role for an individual's knowledge. Undermining the role of any autonomous individual decision maker in turn undermines the purpose for building neoclassical equilibrium models.[51]

50. One is not really free to choose – if we mean by that a *thinking* person making a choice – if one's choice is dictated by one's genetically (and thus exogenously) given 'tastes'.

51. For an extensive discussion of methodological individualism and why we do not need to rely on a psychologistic basis for explaining individual choices, see Boland [1992, ch. 10].

# CHAPTER 7

✧

# Limits of equilibrium methodology

## An educational dialogue

Today, almost every economics professor has had to teach the proverbial Economics 101 (which usually refers to micro rather than macro). Most of the textbooks they use only provide some very basic methodological ideas. The most obvious ideas in North American economics classes are usually about the virtuous nature of market competition and the unvirtuous nature of collusion and monopolies. The least obvious methodological idea was the one discussed in Chapter 6 – methodological individualism – which is an essential basis for determining what is required for an acceptable neoclassical economics explanation of any social event. I have been summarizing methodological individualism as the assertion that says 'things do not decide; only people do'. This requirement is of particular concern today among macroeconomic equilibrium model builders but is rarely discussed explicitly in methodological terms.

There is another essential idea rarely discussed. While textbooks make considerable use of states of equilibrium, there is rarely any discussion of equilibrium stability – as Fisher observed [1976, p. 3]:

> There was a time, about fifteen years ago or so, when the stability of general equilibrium was a hot topic among mathematical economists. . . . [T]here was a flurry of papers . . . Nowadays, however, the subject, if not actually disreputable, is at least not very fashionable. . . . I am convinced that the problems involved in stability analysis are of central importance to economic theorists, and, indeed, to economists generally. The fact that we know so little about the answers should not blind us to the importance of the questions.

Even though a stable market is essential for models purporting to explain prices, unfortunately few teachers or textbooks ever properly characterize equilibria such that they can be distinguished from simple balances.[1] And, I suspect this is the reason few if any textbooks say much about the necessity of assuring equilibrium stability in equilibrium models designed to explain prices. However, some teachers might be much more critical and make an observation like that of the historian Carl Becker: 'Theology in the thirteenth century presented the story of man and the world according to the divine plan of salvation. It provided the men of that age with an authentic philosophy of history, and they could afford to ignore the factual experience of mankind since they were so well assured of its ultimate cause and significance' [1932, p. 17].

## 7.1. A DIALOGUE IN AN ECONOMICS 101 CLASS

Here I will discuss the main problem of what is found in virtually all microeconomics textbooks. I will illustrate it using the following dialogue between an inquisitive student (S) and a typical economics teacher (T).[2]

**S:**   Professor, why is the price of any good (e.g. apples) what it is?
**T:**   Because the observed level of the price for any good is at an equilibrium level – look at Figure 7.1 where the equilibrium price is $P_e$.
**S:**   What do you mean by an 'equilibrium level'?
**T:**   Well, had the price (for any reason) been higher it would fall back down, and had it been lower it would rise back up.
**S:**   Why might this be so?
**T:**   Because it is the nature of any world of rational and price-competitive people. Specifically, it is because:

(1)  *The nature of the world is like this:* either the price, $P$, equals $P_e$ and demand equals supply, or
    (i)   anytime $P > P_e$ there will be 'excess supply' (ES), and
    (ii)  anytime $P < P_e$ there will be 'excess demand' (ED).
(2)  *People are 'rational' which means:*
    (iii) demanders seek to maximize their utility or satisfaction, and
    (iv)  suppliers seek to maximize their profits.

---

1. As I briefly noted in note 10 of Chapter 5.
2. I have borrowed and revised this dialogue from Boland [1986].

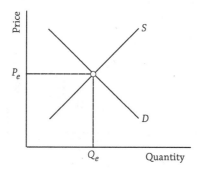

**Figure 7.1.** The equilibrium price

(3) *People are price-competitive:*[3]
   (v) anytime there is excess supply, someone will bid the price down, and
   (vi) anytime there is excess demand, someone will bid the price up.

**S:** For me to understand your claim for the nature of the world, I need to be able to see how the world might conceivably be otherwise. Help me with Figure 7.2 on which I have drawn all six possible configurations of the slopes of demand and supply curves at their intersection (I have left out the special cases involving vertical or horizontal curves or equal slopes to keep things simple). I now see that your claim about the nature of the world and its price adjustment behaviour is that the market must be as shown in Figure 7.2(a), (b), or (c) and thus your claim is really that the world is not as shown in Figure 7.2(d), (e), or (f). This is because, had the world been as shown in Figure 7.2(d), (e), or (f), then the competitive behaviour you claim for people would cause the price to move away from the equilibrium and thus the least likely price to observe would be the 'equilibrium' price, $P_e$. But I fail to see how you or anyone else can distinguish between Figure 7.2(a) and (d) or between Figure 7.2(c) and (f) without violating the methodological individualist view that demanders and suppliers make their autonomous decisions independently. For example, if both curves are downward sloping (perhaps it is in a market in which the sellers give quantity discounts) how do we know the world is like Figure 7.2(a) rather than like Figure 7.2(d) without presuming that the slope of the demanders' demand curve is not in some way constrained by the slope of the sellers' supply curve?

---

3. Competitive *price-adjustment* behaviour is usually called Walrasian behaviour.

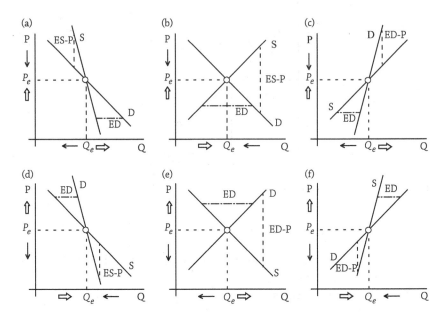

**Figure 7.2.** Possible markets

**T:** You are right. There would seem to be a potential methodological problem here, but there are some disequilibrium responses hidden in our textbook's theory of the individual supplier to take care of this.[4] Specifically, by saying that the individual firm produces where profit is being maximized for the given (demand) price, I am in effect saying that the firm responds to any difference between the going demand-price and the firm's marginal cost, since the marginal cost indicates the profit maximizing 'supply-price'.[5] That is, we can add to the list two implicit behavioural conditions of the profit maximizing behaviour of competitive price-taking firms:[6]

(v') if there is an excess supply-price (P < MC), the firm will decrease the quantity supplied, and
(vi') if there is an excess demand-price (P > MC), the firm will increase the quantity supplied.

4. The teacher here has in mind the textbooks' Marshallian theory of the firm with its rising marginal cost curve.
5. That is, whenever given the market's 'demand-price' (defined at each quantity of demand such that each demander is maximizing utility) the firm will choose the corresponding supply quantity for which profit is maximized. At each level of supply, the marginal cost indicates what the price would have to be for that level of supply to be chosen; hence the corresponding marginal cost equals the supply-price.
6. The following competitive *quantity-adjustment* behaviour is called Marshallian behaviour since it is implied by Marshall's theory of the firm.

If I may use your Figure 7.2 to put things into your terms, we see that the implications of this quantity-adjustment behaviour are that my claim about the nature of the world is that the profit maximizing behaviour of the firm additionally means that the world cannot be like Figure 7.2(a), (c), or (e) since the quantity-adjustment behaviour of our textbook's firm would move the quantity away from the equilibrium quantity, $Q_e$. Thus, if people behave as I claim (as profit maximizers *and* price-competitors), then only Figure 7.2(b) could ever represent the possible world. So, the potential methodological problem that was bothering you disappears. My claim boils down to one that the world is like Figure 7.2(b) and in such a world profit maximizing with price-competitive behaviour will always lead to equilibrium prices (and equilibrium quantities, too). And, given this necessary tendency towards states of equilibrium and the ability to show mathematically that any state of equilibrium can be shown to imply a Pareto optimum[7] with respect to resource allocations, you can see why I am trying to convince you that we should all put a price on our services and get out there and compete in the market. If we all do so, it will be the 'best of all possible worlds'.

**S:** Not so fast, I am not so convinced. Although it might be conceivable that people can be so competitive, why might the world necessarily be only like Figure 7.2(b)?

**T:** Well, you can see that Figure 7.2(b) has very convenient properties. If we can show that all demand curves are downward sloping as a consequence of consumers' independent and autonomous decision-situations and that all supply curves are upward sloping as a consequence of the firms' independent and autonomous decision-situations, then the requirements of methodological individualism are satisfied. Before you jump on me to say that these are market curves and not individual curves, let me say that I can provide the mathematics to show this for each individual consumer's demand and each individual producer's supply curve. Thus, if every individual's demand curve is downward sloping, then necessarily when I add together all of the consumer's individually demanded quantities at each price, the resulting market curve will be downward sloping. The related case is true for the sum of all of the firm's individual supply quantities at each price – that is, the resulting market supply curve will necessarily be upward sloping. The key issue that must be stressed here is that there is to be no need for collusion between decision-makers when they make their autonomous decision independently. ʹ

7. For those unfamiliar with the jargon, a Pareto optimum occurs in an economy when no one person can gain unless someone else loses.

**S:** Leaving the question of collusion aside, you have not answered my question. You have told me what you want – a situation where everyone can be independent and still have the possibility that the unintended consequence is an equilibrium with all its Pareto efficiency benefits – but you have not yet told me why the world *is* like Figure 7.2(b).

**T:** Very well, but you may still not be convinced. The reason all individual demand curves are downward sloping is because, psychologically, all individuals face given utility functions which have the common property that marginal utility is always diminishing. And supply curves are always rising because every firm's productive capabilities on the margin must be diminishing. These two concepts of diminishing margins are facts of nature and when combined with 'rational' decision-making (constrained optimization) will necessarily lead to the proper slopes as shown in Figure 7.2(b).

**S:** I am still not convinced since I read in our textbook that diminishing marginal utility is not sufficiently limiting because with it all that we can say is that for a demand curve to be rising the good must be an inferior good (i.e., a good which one will buy less of when one's income increases). Is there something more to your claim about the nature of the world?

**T:** Does there have to be more? Even if some individuals have upward sloping demand curves, it is unlikely that all do and so when we add up their respective demand quantities we will find that the aggregate market curve is still downward sloping. In effect, if there are just a few perverse people, their behaviour will be cancelled out by the dominant behaviour of normal people.[8]

**S:** Now again you seem to be going for your convenience rather than trying to convince me about the nature of the world and why I should eagerly want to engage in the competitive market system. If some consumers have upward sloping curves and some have downward sloping curves, where do we draw the line? It is certainly possible that the number of people with downward sloping curves is just about enough to be cancelled out by the number with upward sloping curves. So, for the last person whose demand is to be added to the market total, how do we avoid violating methodological individualism if we have to exogenously constrain the last individual to have a downward sloping demand curve (to preserve the negative sloping market demand curve)?

**T:** I am beginning to wonder who the teacher is here since many of your questions are longer than my answers. You seem to be suggesting that either I violate methodological individualism in order to convince you

---

8. If the teacher thought there were sufficient time, I am sure he or she here would have quoted Hicks [1956, pp. 67 and 92–4], who argues this. However, I doubt he or she would have bothered to mention my counterargument in Boland [1992, ch. 14].

that competition is a good thing or I go back to the drawing board to see if I can come up with a separate way of ensuring the stability of the market without violating methodological individualism, perhaps by showing why an individual's method of responding to disequilibrium situations guarantees stability.[9]

**S:** Well, the latter would certainly help. But, I must caution you that I will not be satisfied unless the separate way assuredly does not violate methodological individualism. That is, it must only be determined by exogenous factors that are psychologically or naturally given. At the very minimum, I would like to know how individuals learn to respond in such a stabilizing manner.

**T:** Now you have reached an easy question. Let us go back to the beginning and do it all over and I am sure you will see how this question is answered.[10]

## 7.2. THE STABILITY IN AN EQUILIBRIUM MODEL MUST BE ENDOGENOUS

In this dialogue we see the seeds of many research programs – the most obvious is the one suggested by Arrow in his 1959 article discussed in Chapter 2. The inherent stability of the textbooks' neoclassical model corresponding to Figure 7.2(b) is clearly necessary for the normative conclusions often promoted in economics classes. Yet as the dialogue illustrates, the logic of individual decision-making does not by itself ensure that only Figure 7.2(b) is the true representation of the real world. That is to say, the stability of the market

9. At this point if the teacher is a member of the older culture (the non-formalist critics discussed in Chapter 5) rather than the newer formal mathematical model builder culture, he or she might think this has all been solved years ago by the consideration of the cobweb theorem about the so-called hog cycle (which I briefly discussed in Chapter 5). Unfortunately, the hog cycle is a market situation in which there is a sequence of alternating quantity and price adjustments. The so-called stability assured by the cobweb theorem (which requires that the slope of the supply curve be steeper than the slope of the demand curve) is not what the student here is talking about nor is it what formal mathematical equilibrium models are talking about. Moreover, such a requirement would violate the methodological individualism requirement of neoclassical equilibrium models since it makes the behaviour of the demanders and their market demand curve depend on the behaviour of the suppliers and their market supply curve. For a fuller discussion of the failure of the cobweb theorem to yield stability, see Pashigian [1987] who also points to Muth [1961], who criticized the cobweb theorem's notion of stability.

10. At this point if the student had paid attention in his or her history class and read the small book by Carl Becker [1932] quoted at the top of this chapter, he or she might ask whether this is a thirteenth century economics class.

is not obviously endogenous. Specifically, if it is possible for Figure 7.2(a), (c), (d), (e), or (f) to be true representations, there must be another way to ensure stability beyond analytically specifying mechanical responses to (positive or negative) excess demands or excess demand-prices beyond that suggested in Arrow's 1959 article. A not-so-obvious alternative is to explain the stability as an outcome of the learning process which is implicit in the recognition that every decision-maker's knowledge of the decision situation is limited and thus the correct expectations must be learned as part of the process of reaching the equilibrium as implied by Clower's 1959 Model III of the ignorant monopolist. A too-obvious alternative is to ignore the difficulties of the microeconomic behaviour and revert to the analysis of aggregates or of DSGE models[11] and thereby avoid the complexities of the questions of endogenous stability. Surely, this might be seen as a way to avoid the difficulties of assuring equilibrium attainment identified by Richardson in his 1959 article.

But, I think the real import of the dialogue is that there is something quite unconvincing about what we teach in Economics 101. What is most unconvincing is that if one has to assume the existence of a state of equilibrium to make any policy point whatsoever, the realism of the policy point will at least be open to question. Economics 101 taught as the dialogue's economics Teacher tries is dangerous and misleading if that is the only basis for today's students' understanding of economics and for those who might go on to be politicians. After all, Economics 101 is probably most politicians' only basis for understanding the economy. Competition may be a good thing, surely, but it cannot be just because of the properties of a possible – or at most conceivable – state of equilibrium. Clearly judging by the above dialogue, just assuming the existence of a state of long- or short-run equilibrium is almost always open to the criticism of being very unrealistic.

11. These models will be discussed in Chapter 10.

# CHAPTER 8

꿰

# Equilibrium models vs. realistic understanding

A s I discussed at the beginning of Chapter 6, Austrian critics of neoclassical economics have for decades called attention to the knowledge requirements for any state of equilibrium.[1] But Lachmann went even further [1982, p. 636]:

> After what has happened in economics in the last 30 years we are today inclined to look askance at the whole notion of equilibrium, and even more so at the Hayekian version of 1936 in which we were told 'It can hardly mean anything but that, under certain conditions, the knowledge and intentions of the different members of society are supposed to come more and more into agreement'. . . . But even if we discard the equilibrium terms in which the problem was first stated, it nevertheless remains. In a stationary world . . . time will in the long run, 'hammer logic into brains' and teach its human pupils what they must do to achieve success and stave off failure. Why this should be so in a changing world is by no means clear.

In this chapter I will be arguing that the basis for discussing such 'knowledge requirements' usually turns out to be a presumed theory of knowledge that is untenable.[2] The important question is whether *any* theory of knowledge could be consistent with the knowledge requirements of a state of equilibrium.[3] This

---

1. For example, see Shackle [1972] and Lachmann [1976, 1982].
2. I discussed this in Chapter 6 and will be discussing knowledge also in Part III. For even more about this, see Boland [2003, ch. 10; 1997, ch. 6].
3. For even more discussion specifically about knowledge in neoclassical economics, see Boland [1992, ch. 6].

question is not insignificant as there have been strong claims made that the information contained in any set of equilibrium prices is complete.[4] There is a related and more fundamental question that must be considered – one that addresses the dynamics required when recognizing knowledge in equilibrium models. Is the *process* of acquiring the necessary knowledge consistent with the logical requirements for a state of equilibrium? Perhaps, if the concept of equilibrium is properly specified, the necessary information for convergence to equilibrium can be provided automatically in any state of disequilibrium as Hayek suggested in his 1945 article. These questions of a realistic relationship between the equilibrium process and a learning process that would be consistent with a state of equilibrium will be examined in this chapter.

## 8.1. EQUILIBRIUM ATTAINMENT AND KNOWLEDGE SUFFICIENCY

Tjalling Koopmans was a prominent member of the newer culture that promoted formal, mathematical equilibrium model building that I discussed in Chapter 5.[5] Koopmans published an important book in 1957 that promoted the, then, common view of mathematical equilibrium model building. In it he explained [1957, p. 53]:

> The [equilibrium] price system carries to each producer, resource holder, or consumer a summary of information about the production possibilities, resource availabilities and preferences of all other decision makers. Under the conditions postulated, this summary is all that is needed to keep all decision makers reconciled with a Pareto optimal state once it has been established.

Here, his talk about postulated conditions is just about the ordinary equilibrium conditions such as those identified in Arrow's 1959 article. But two questions need to be asked. First, how does the autonomous individual decision-maker acquire the information comprising the *whole* 'price system'? Second, does the process of establishing the Pareto-optimal state require information different from that which reconciles the independent decision making of individuals? Before considering these questions, the role of the information contained in an equilibrium price system with respect to an individual's decision process needs to be examined.

---

4. For example, see Hayek [1945, p. 526] and Koopmans [1957, p. 53].
5. In the remainder of this chapter I will be revisiting some of the questions I raised in Boland [1986].

Obviously, if we ignore realism[6] as early mathematical model builders such as Koopmans were willing to do, and if we are willing to assume every individual knows the equilibrium price for every good that is considered, and to assume every individual actually faces those equilibrium prices, then maximizing behaviour cannot yield a disequilibrium. But is knowledge of the equilibrium price system all that is required? Should we not also have some knowledge of the availability of the goods demanded? If we are really discussing an individual's demand decision at the market equilibrium point and all individuals are facing the same price, the supply will be just enough to meet everyone's demand. Viewed this way – that is, looking only at the quantities demanded and supplied at the equilibrium point where each individual knows all the relevant prices – as most Economics 101 textbooks would suggest – there cannot be a problem of availabilities, since the additional knowledge of availabilities is redundant.

The question of the sufficiency of the knowledge of the equilibrium price system is obviously important. Is the number of equilibrium prices that the decision-maker needs to know more than could be considered reasonable or even realistic? And, how does any ordinary individual know that those prices are the equilibrium prices? Whether the number of goods for which the individual must have information is unreasonable depends on the specific equilibrium model under consideration. Clearly a model in which it is assumed that there is an unlimited number of goods contains an assumption which would put a considerable strain on the credibility of Koopmans' claim. Perhaps a more modest model with a small number of goods might seem reasonable. What seems reasonable will depend on the theory of knowledge presumed to apply to the individuals in question. I will return to this issue a little later. For now, let us assume the number is reasonable and proceed to another question.

The question of *how* an individual knows that the given prices are equilibrium prices is in one sense beside the point. As I have been noting, if they are equilibrium prices, any individual's planned purchase pattern will be fulfilled.[7] But if the individual cannot be certain that the prices faced are equilibrium prices, why should anyone expect that the planned purchase pattern would be the same as the plan formed when the prices are certainly equilibrium prices?

Consider a simple decision situation facing the usual textbooks' consumer. If we recognize that any individual cannot be in two places at the same time – say two shops in the same shopping mall – then the individual must decide which shop to go to first (e.g., the butcher or the baker). Obviously, if you

6. Critics who complain about the realism of a neoclassical equilibrium model are usually complaining that there is no assurance that the assumptions of such an equilibrium model are true – that is, that they exactly fit the facts.

7. As I am still only talking about a competitive market equilibrium with all participants being price takers, this will be true.

thought there might be a shortage at a particular shop you might want to plan to go to that shop first. In this sense, knowledge of the equilibrium status of the price system is essential. If the individual does not know that the prices are equilibrium prices, his or her plan will be a bit more complicated. Nevertheless, if the prices are the equilibrium prices, the extra complications should not matter since all aspects of the plan are fulfilled in the end.

If we limit our consideration of the economy to equilibrium price systems, certain social liberal ideological implications follow.[8] If the decision-makers all face the same equilibrium price system, then their independent and autonomous decisions are perfectly coordinated (i.e., demand equals supply in every market) and it does not matter how the prices were established. Koopmans' definition of an equilibrium price system 'does not necessarily presuppose the existence of a competitive market organization' [1957, pp. 50–1]. And he says [p. 53], 'Discussions of pricing as a tool for planning and operating a social-ist economy likewise derive from our proposition [5]'.[9] So, if we restrict our discussion to equilibrium price systems, we do not have to be concerned with whether competition as a process is a necessary, or even a good, thing. This is optimistic social liberalism at its best, but it sure misses the point of why one would ever want to argue in favour of market competition following the tradi-tion promoted by Adam Smith.

Judging by these quotations from Koopmans' works, some may find it interesting that while Koopmans and Hayek seem to be in complete agree-ment concerning the informational efficiency of an equilibrium market price system, the ideological perspectives implicit in their respective views of equi-librium models are usually considered less compatible.[10] In his 1945 article Hayek argued that only the competitive market price system is efficient, and even when based on prices, the socialist planning system is virtually impossi-ble, let alone efficient. While Koopmans' view of the price system might seem to promote some form of social liberalism at least when it comes to the ques-tion of the viability of socialism or socialist planning, Hayek's would seem to

8. I am using the term 'social liberal' to identify what is just called 'liberal' or 'pro-gressive' in North America and perhaps called social-democratic elsewhere. I am using this term to distinguish it from the neo-liberal ideology which is considered 'conserva-tive' in North America – particularly when it comes to the belief that the market can solve all problems. Note also that I am using 'liberal' not 'libertarian'. I use liberal here only to mean that there is a liberal tolerance for a role of government or even for a limited degree of social planning.

9. No need to discuss this proposition here other than note that his 'Proposition 4' was a proof that every competitive general equilibrium is also a Pareto optimum and his 'Proposition 5' was an attempt to prove that for any Pareto optimum one can define a competitive general equilibrium set of prices.

10. Note however, Koopmans seems to have kept his actual political and ideologi-cal views to himself so one can only infer his views from those statements I have been quoting.

be based on a conservative or neo-liberal view of the nature of the competitive market system.[11] If Hayek had focused his view of the informational efficiency of the price system on the logical properties of a state of equilibrium, his view would be difficult to sustain for the following reason. In a truly competitive (long-run) equilibrium all production functions, of mathematical necessity, must be locally linear-homogeneous (i.e., all exhibit constant returns to scale on the margin[12]) which means a labour-theory of value mathematically yields the same conclusions concerning income distribution as does any other theory of value (such as a capital-theory of value).[13]

A formal competitive equilibrium model is truly the domain of social-liberal economists. While one might agree that an open-minded social liberalism is admirable, it is risky to base it only on the properties of conceptually narrow and possibly unrealistic equilibrium models. But this can also be true of the conservative or neo-liberal view that sees many virtues in a perfectly competitive general equilibrium model that economists such as Hayek promoted in his 1945 article.

---

11. In North America this is a conservative ideology only in the sense of advocating limiting the role of government in the market economy.

12. For those unfamiliar with what is taught in beginning textbooks, a price-taking perfectly competitive firm in a model is said to be in a long-run equilibrium whenever enough time is allowed for that firm to make optimal choices concerning all inputs as required for all states of equilibrium. In such a long-run equilibrium the firm is choosing to produce at a level of output by choosing its labour input such that profit is being maximized and unintentionally the average cost for each unit of output is at its minimum at that point. This turns out to mean that the price is equal to both the average and marginal costs – that is, equal to the average total cost of all inputs and to the extra cost that would occur if one more unit of output were produced. As I explained in Chapter 3, the firm is usually assumed to be facing a U-shaped average cost curve (one for which average cost is initially falling but eventually begins rising as more is produced, hence the U-shape) such that the point at the bottom of the average cost curve corresponds to a point on the production function that is (at least very nearby or 'locally') mathematically linear and homogenous (implying that at that point the average cost is temporarily proportional to output). For more on this consult the chapter about cost curves of any beginning microeconomics textbook.

13. For example, consider again the simple model of the Prologue in which a firm's total cost (TC) measured in dollars is the sum of two terms such that $TC = W{\cdot}L + P_K{\cdot}K$. These two terms are what has been paid to the suppliers of the two inputs, $L$ and $K$, given their dollar market prices $W$ and $P_K$. These terms, in effect, represent incomes of the respective input suppliers. Here TC is being measured in dollars but it could also be measured in units of any non-price variable if that variable's price is known – it is just a matter of algebraic bookkeeping. Here we could measure the total cost in units of labour as $TC_L = L + (P_K/W){\cdot}K$ or in units of capital as $TC_K = (W/P_K){\cdot}L + K$. This will be true for any firm in a state of *long-run* equilibrium. I raise all this because many neoclassical economists in the 1960s and 1970s associated the labour theory of value with Marxian socialist economics. But if we only look at firms in states of long-run equilibrium, there is no such difference.

## 8.2. EQUILIBRIUM MODELS AND
## THE IGNORANT CONSUMER

Consider again an independent individual who knows the going market prices for the goods that he or she wishes to purchase but this time does not know whether they are equilibrium prices and hence cannot be sure that there will be enough of everything nor whether, when some market does not clear, the price will rise or fall. If equilibrium model builders are going to build a model of the process of reaching an equilibrium, they need to consider how they should characterize such an individual's decision process. Of course, it could be assumed that the individual blindly flips a coin to decide whether to go first to the butcher or to the baker. That would not be much of an explanation. So, let us consider the model builder following the lead of Hayek, Fisher, or Hicks and see the individual forming a 'plan'. To do this, model builders need to specify the essential elements of a plan, particularly if they are to consider what some might call an 'optimal plan'. Most importantly, they must be careful not to predispose the conception of a plan to be consistent only with the requirements of a state of equilibrium (as Hayek did in his 1937 and 1945 articles) for that would make their model's explanation circular.

Let us return to consider a single individual visiting the two shops in the shopping mall and use Figure 8.1 to focus on the plan of this individual. To be in accordance with the traditional textbooks' theory I will, of course, simply need to assume that the individual knows his or her indifference map and

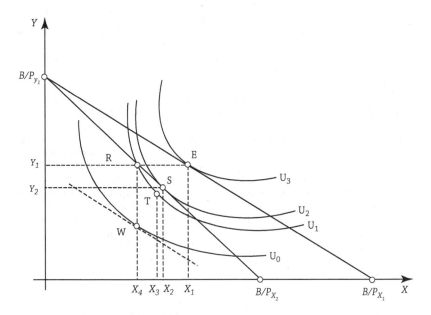

**Figure 8.1.** Choice facing unavailability

budget, $B$, that are illustrated in this figure. For this individual's situation, what will definitely be in doubt is this individual's knowledge of the prices, $P_x$ and $P_y$, and possibly the eventual availability of the goods. That is to say, for this individual, the preference map and budget shown in Figure 8.1 just illustrates this individual's *expectations* of the decision situation being faced.

Consider again the individual's knowledge before entering the shopping mall and let us ask about what the individual in his or her ignorance does to form expectations. As I discussed in Chapter 4 and as Richardson suggested in his 1959 article, prices depend on the behaviour of all other individuals in the market, so it seems reasonable to think that it is unlikely for an independent individual to know what all other individuals will demand or supply and thereby to use such knowledge to calculate equilibrium prices. For now I will ignore what I discussed and questioned in Chapter 6 about it usually being taken for granted in textbooks that the individual knows his or her preference map. But some might still question allowing for the a priori knowledge of the budget if the latter depends on income to be earned in, say, the labour market since the labour market may not be in equilibrium.[14] But to keep things simple let us just maintain the textbooks' view that the budget, $B$, and the preference map are known.

The ability of individuals to form correct expectations of the equilibrium prices would usually be the focus of any theory of an optimal plan. Consider again Figure 8.1: if the individual expects the prices to be $P_{x_1}$ and $P_{y_1}$, the plan would be to go to the respective shops and buy point E which represents quantities $X_1$ and $Y_1$. If the individual accidentally forms expectations of prices that happen to correspond to the equilibrium prices, these quantities will be successfully purchased. However, if the expected prices are wrong, the optimum point in the (*ex ante*) optimum plan, E, will not be the (*ex post*) optimum once the actual prices are known. How does the discrepancy between the expected situation and the actual one manifest itself? Again, this is a matter of the individual strategically having a plan in mind. And unlike Clower's ignorant monopolist I discussed in Chapter 3, this consumer's plan must recognize the possibility of erroneous expectations. Many strategies are possible, but for now let us just say that the individual thinks that if either of the two shops is to be short of supply it is likely to be the shop for good $Y$ and thus the optimum plan is to go to that shop first.[15] Let us further say that the individual successfully purchases the planned amount of $Y$, $Y_1$, at the expected price, but when the individual later arrives at the shop for $X$, things are not as expected.

14. This is discussed by Clower [1965] who observes that for there to be an equilibrium in the market for a consumer commodity, the consumers must also be in equilibrium in the markets where their budgets are financed as in the case of the labour market.

15. As I suggested earlier, think of this as choosing between the butcher and the baker.

In this situation there are two different ways the consumer's expectations could be wrong: There is the expected-price error when the actual price is not the expected price and there is the expected-quantity error when the actual price is as expected but the supply is not sufficient. Recognizing errors as expected-price errors is consistent with the textbooks' theory of perfect competition in which the market is assumed to be very large such that each individual is considered relatively insignificant and thus has no effect on any shop's market price. It is also consistent with the theories in which the price responds instantly to shortages so that at the actual shop's clearing price there is a sufficient supply.[16] But considering all errors to be expected-price errors may be predisposing the view of the plan to be one which is consistent only with an equilibrium model. Since the task here is to model the process of reaching a market equilibrium, both types of expectational error need to be addressed.

Figure 8.1 can be used to illustrate the differences in the types of error. The expected-price error is viewed as follows. If the individual has already purchased $Y_1$ amount of good Y, but the actual price of X is $P_{X2}$ and not the expected $P_{X1}$, when he or she arrives at the shop for X, the planned optimum, $X_1$ is outside the actual budget line. The best that can be done[17] is to buy the amount $X_4$, which is not only less than the planned optimum, but is also less than the textbooks' optimum for the actual prices.[18] That is, if the individual knew in advance the actual prices, he or she would have wanted to purchase point S, which represents $Y_2$ and $X_2$; but this is not possible since, as I said, the individual has already purchased $Y_1$. Here we would say the individual buys $X_4$ because the actual price of X was not expected and point R is the best that can now be done.

Consider now the expected-quantity error. Let us say prices are as expected ($P_{X1}$ and $P_{Y1}$) and at the second shop the individual buys $X_4$ because that happens to be all that was available on the shelf. (Of course, this may only beg the question about why this individual was the last one to find anything on the shelf.[19]) Under these circumstances, $X_4$ is still the best that can be done, even though the prices may be as expected. But if the actual prices are the expected $P_{X1}$ and $P_{Y1}$, given that the consumer purchased $Y_1$, the textbooks' optimizing

16. Obviously, the shops for Figure 8.1 are not Walrasian general equilibrium markets where transactions are not allowed until everyone is maximizing.

17. Where the best is the highest possible level of satisfaction represented by an indifference curve intersecting the line representing the already purchased level of Y.

18. The textbook optimum is always where the marginal rate of substitution (the slope of the indifference curve) equals the slope of the budget line (which is the relative price, $P_X/P_Y$).

19. But here it is simply because transactions were being allowed prior to any equilibrium.

point for a prior choice of $X_4$ would be at $W$.[20] Point $T$ may be a more efficient way to achieve the same level of utility as point $R$, but it still requires more $X$ than can be purchased after $Y_1$ has been purchased first – as before, the optimum point is not possible if the individual has already bought the planned amount of $Y$ at the first visited shop. From either perspective (expected-price or expected-quantity errors), according to either the textbooks' calculus optimization rule or the consumer's expectations, the individual is not optimizing at point R. Although, given the expectations concerning availability, the individual's optimum plan said to go to $Y$'s market first, it turned out that it was the expectations about availability that may have been erroneous.

Consideration of what I have illustrated in Figure 8.1 highlights a major concern of economic model builders who wish to recognize disequilibria without giving up equilibrium models. How does the individual become aware that the market is not in equilibrium? That is, is it a matter of expected-price errors or expected-quantity errors? Unfortunately, recognizing 'disequilibrium awareness' in the model is not enough. The model builder must also explain how the individual learns to respond in a manner that promotes a movement toward equilibrium and thereby ensures the future ability to fulfil consumption plans. In other words, we must explain how the individual learns to form more accurate expectations.[21]

## 8.3. THE MARKET'S EQUILIBRIUM PRICE: LEARNING VS. KNOWING

Admittedly it is a difficult and complex task to provide a microeconomics explanation of disequilibrium behaviour that is consistent with the explanation of equilibrium behaviour.[22] Besides retreating to macroeconomics (as will be discussed in the next chapter), there is another way to avoid the complexities. A model builder could give up any reliance on an assumption of an already existing state of equilibrium or on an assumption that all prices are equilibrium prices. This would seem to be the obvious advantage of Hayek's earlier writings, which stressed the need to recognize the role of information and knowledge before reaching a state of equilibrium. For him it is the optimality of the equilibrium *process* that shows the virtues of the competitive price system. If each individual could be seen to respond to the failures of just his or

20. Optimizing again only in the sense that the necessary condition for utility maximization (equal slopes) is obtained given the prior purchase of $Y_1$.

21. In Chapter 9 I will be addressing how some macroeconomic equilibrium model builders are viewing expectation formation today.

22. Of course, this was evident in Chapter 2's discussion of Arrow's 1959 article in which he suggested two different and conflicting theories of price determination – one for when the market is in equilibrium and another for when it is not.

her personal unfulfilled plans – without having to consider what anyone else is doing – one might be able to agree with Hayek's view of the competitive price system. Moreover, for Hayek an individual needs only to learn about the going prices and his or her personal situation. In this regard, one could optimistically agree with the view that the methodological-individualist conception of free-enterprise capitalism would necessarily have a distinct advantage over any social organization that might be based on the perspective of a socialist planner. But as is apparent in Koopmans' optimistic viewpoint, the socialist planner would just have to calculate the equilibrium prices in advance using a formal equilibrium model. For Hayek, though, that socialist planner would need to know about the situations facing every consumer and every firm in the economy.[23]

Hayek's argument is against the possibility of any informationally adequate general equilibrium model which would take the mathematical existence of equilibrium as its central methodological concern.[24] Moreover, this is the crux of his argument for a significant role for information and learning in any competitive equilibrium system. From his perspective, understanding economics is not a matter of a rigorous examination of the mathematical properties of a model's state of equilibrium, but rather it is merely an appreciation of the equilibrium process as being one that always points in the direction of the equilibrium. As a matter of theoretical convenience, Hayek's view has found a certain degree of acceptance[25] as it seems to deal directly with the relationship between learning and the equilibrium process. This relationship is recognized also as essential for the disequilibrium foundations of equilibrium economics. Unfortunately, Hayek's emphasis on studying the learning inherent in the equilibrium process – rather than just the knowledge requirements for any claimed state of equilibrium – relies too much on a questionable presumption which amounts to assuming exogenous stability. Referring back to Chapter 7, consider again Figure 7.2. Only when the true world is as represented by Figure 7.2(b) can the followers of Hayek's view be confident that the individual consumer or firm is learning to respond in the correct way, a way that will lead to a better allocation of resources. This is especially so whenever we give up basing our economic explanations on an assumption that eventually the optimum allocation is *always* achieved.

23. Interestingly, Hayek was talking about planning in the 1940s when such data was not readily available. Today, we have the internet and high-speed computers which might undermine or at least weaken his whole argument.
24. For examples see Wald [1936/51] and Koopmans [1957].
25. As can be seen on page 225 of Richardson's 1959 article.

# CHAPTER 9

ᏅᎧ

# Macroeconomic equilibrium model building and the stability problem

The question of the stability of any assumed equilibrium in a microeconomic model is too often ignored today (as I have repeatedly noted in other chapters). I think it is now time to consider all major ways of either avoiding or addressing the methodological questions posed by stability analysis. The main question concerns how to provide an explanation of disequilibria that is both consistent with any view of the state of equilibrium and faithful to the usual neoclassical commitment to methodological individualism. To begin, consider the view of George Evans and Seppo Honkapohja [2009, p. 421]:

> Expectations play a central role in modern macroeconomics. Economic agents are assumed to be dynamic optimizers whose current economic decisions are the first stage of a dynamic plan. Thus households must be concerned with expected future incomes, employment, inflation, and taxes, as well as the expected trajectory of the stock market and the housing market. Firms must forecast the level of future product demand, wage costs, productivity levels, and foreign exchange rates. Monetary and fiscal policy-makers must forecast inflation and aggregate economic activity and consider both the direct impact of their policies and the indirect effect of policy rules on private-sector expectations.

As can be seen in this quotation as well as in Richardson's 1959 article, general equilibrium attainment is open to question and problematic unless model builders address problems of forming expectations rationally or forming what Richardson called rational expectations. Today, it is macroeconomic model builders who see the need to address rational expectations as a way to address the matter of equilibrium attainment. However, while what Richardson meant by 'rational expectations' in 1959 did not involve any specific way to form expectations, today macroeconomic equilibrium model builders adopt various sophisticated methods using the Rational Expectations Hypothesis introduced by John Muth in his 1961 article about equilibrium attainment in the cobweb model.[1] There are a few problems with relying on rational expectations for equilibrium attainment and they will be discussed in this chapter.

## 9.1. RATIONAL EXPECTATIONS AND MACROECONOMIC EQUILIBRIUM MODELS

Starting in the 1970s, macroeconomic equilibrium model building and its approach to the questions of stability analysis centred on the role of rational expectations.[2] As I have been noting, a significant role for a concept of an individual's decision plan in an equilibrium model would require recognizing his or her formation of expectations. As can be seen in recent literature, how to model the convergence to an equilibrium continues to be of considerable interest.[3] Fisher observed [2003, p. 75]:

> [T]he statement that agents will eventually learn about and act on systematic profit opportunities is an appealing assumption. The proposition of the rational expectations literature that agents always instantaneously understand the opportunities thrown up by an immensely complex and changing economy is breathtakingly stronger. That proposition begs the question of how agents learn . . .

In his 2008 *Palgrave Dictionary* entry about recent developments in general equilibrium modelling, William Zame adds:

> A criticism of rational expectations equilibrium is that it does not address the mechanism through which agents obtain their private information. This seems an important omission because, if all information were to be revealed by prices, there would seem to be no incentive for agents to acquire information in the first place, especially if acquiring information is costly . . . A second criticism

1. I discussed the cobweb model in note 9 of Chapter 7; the Rational Expectation Hypothesis will be explained later.
2. Parts of the remainder of this chapter develop and update arguments made in Boland [1986].
3. I will continue avoiding technical discussion of the matters at hand but those interested in recent attempts to deal with learning in the context of rational expectations might wish to examine some of the recent works some of which I have been quoting. These works might include Sargent [2008], Blume and Easley [1998, 2006], Brunnermeier and Parker [2005] and Hart and Mas-Colell [2003].

of rational expectations equilibrium is that extracting information from prices seems to require agents to have a great deal of information about the economy (including information about other agents) . . . Perhaps the most serious criticism of rational expectations equilibrium is that it provides no process by which information gets into prices. If agents use information in prices in forming their demands, how do those demands influence prices? If demands do not influence prices, where do prices come from?

Earlier, Blume and Easley complained, 'It is inconceivable that individuals should be born with the complete understanding of the economy required by rational expectations equilibria . . .' [1998, pp. 61–2]. Moreover, as Sergiu Hart and Andreu Mas-Colell observe, 'It is notoriously difficult to formulate sensible adaptive dynamics that guarantee convergence to Nash equilibrium.[4] In fact, short of variants of exhaustive search (deterministic or stochastic), there are no general results' [2003, p. 1830]. And as well, we must keep in mind that to be consistent with methodological individualism, any explanation of a convergence to a state of equilibrium must include the explanation of the individual's choices and thus must deal with *how* the individual 'rationally' forms his or her expectations.

What economists mean by 'rational' is not always clear since they use the term interchangeably with 'optimizing' and 'maximizing' and, as I will explain, they usually rely directly or indirectly on a false theory of learning.[5] Unfortunately, the misuse of 'rational' is a common mistake and one that economists continue to make.[6] The main critical point is that rationality is not about some psychological phenomenon – instead it is about the logical validity of an argument. The rationality of an argument ensures that any two

---

4. For readers unfamiliar with game theory jargon, a Nash equilibrium merely identifies an equilibrium situation where every game player is making their personal optimum choice given that all other players are also making their personal optimum choices and so nobody has a reason to change.

5. As will be seen later, many macroeconomic equilibrium model builders identify the formation of expectations that are correct on average as being rational. This is just their definition of optimum and not a different notion about rational.

6. An argument could be made that economists make this mistake because economics as a subject for study began in the eighteenth century when rationality was promoted by those 'rationalist' philosophers who saw inductive rational argument alone as a clear means to avoid conceding any authority to the Church. If one were 'rational' one did not have to seek approval of the Church whenever one could prove the truth of one's propositions with observations alone. Carl Becker's 1932 book, *The Heavenly City of the Eighteenth-Century Philosophers*, is a book all economists need to read to get a handle on this. When I was a schoolkid we were all told that what distinguishes animals from humans was that the latter could be rational and the former could not – I suspect we were told this so as to perpetuate this eighteenth-century notion of rationality although those saying this probably were unaware of what they were promoting.

individuals who start from the same premises will reach the same conclusions – this is what the logical validity of an argument assures.[7] Optimality or maximization in any decision-making situation can ensure an equivalent agreement about outcomes. Consider, for example, two textbook consumers whose utility functions are exactly the same function, who face the same prices and face the same budget constraint. Obviously, given their identical decision situations, the 'rational' choices they make will be exactly the same whenever their choices are optimizing or are maximizing. Note, however, that optimality or maximization implies rationality in this sense,[8] but rationality itself need not imply optimality or maximization. The term 'rational expectations'[9] today is used to indicate that if any two individuals are forming rational expectations, they will form the same expectations whenever they face exactly the same information.[10] But, does this mean that rational expectations are necessarily optimal expectations?

### 9.1.1. Rational expectations in a microeconomic context

Before discussing the use of rational expectations in macroeconomic equilibrium models, I need to discuss how rational expectations might be seen in a microeconomic context, in order to be able to address the neoclassical requirement of methodological individualism. The fundamental question is about how individual decision makers learn or form expectations. This may be two separate questions. One question concerns theories about how people learn from objective information and thus about the extent to which observations (or 'data') matter. The other deals with the adequacy of the 'information set' (i.e., the collection of observational data) for the formation of any expectations.

The matter of explaining how individual decision makers form their expectations is frequently recognized, but unfortunately not much has been accomplished at the level of microeconomics beyond what Muth posited in 1961.[11]

---

7. Namely, *if* one's assumptions are *all* true and one's argument is logically valid, *then* one's conclusions *must* be true.

8. That is, to claim a consumer is 'rational' simply and only means one can form a rational argument that thereby proves that his or her utility is maximized.

9. Specifically, the term more commonly used today than the term Richardson used in his 1959 article.

10. Allowing for acceptable stochastic variation, of course.

11. For an excellent history of economic thought article about Muth's creation and use of what he called Rational Expectations, see Sent [2002]. There she observes 'Muth developed the hypothesis mostly in the context of microeconomics, restricting his attention to a single market in partial equilibrium, and had nothing to do with its elaboration in macroeconomics' [p. 305]. This is significant because his hypothesis has been employed almost exclusively in macroeconomic equilibrium models.

Today, it is mainly macroeconomic model builders who employ Muth's posited 'Rational Expectation Hypothesis'. And as to the applications of his hypothesis to just macroeconomic models, it is interesting to note that Muth himself questioned this [1987, p. 97]:

> The application of rational expectations primarily to macro-economics has been a source of amusement to me because I do not now, and never have, understood macro-economics. It has always seemed to be 'half obvious, half un-understandable'. The work in expectations has taken a customarily doctrinaire stance: comparing naïve or exponentially weighted moving averages with rational expectations. There has been little work on developing other hypotheses, particularly those which recognize known cognitive biases in human decision making.

Nevertheless, some model builders still need to address how individual decision makers learn or form expectations. A common way is for a model builder to just assume that there is one sure method of learning from the observations of the economy and that every individual chooses to be guided by this method. This is one way to avoid any complexities of having to model each individual's learning experience. Alternatively, the model builder could just examine the nature of the universal information set since it is assumed to be the ultimate basis of every individual's expectations. Supposedly, how people actually learn from the information set will not matter so long as their learning method is 'rational'.

In order to employ such a position regarding expectation formation, the microeconomic equilibrium model builder would have to rely on two key questionable methodological assumptions. First, and foremost, is the assumption that supposedly there is one and only one universal method of learning. Second, and almost as important, is the assumption that the method employed is a sure method: if we compare the expectations formed from two information sets, and if the two information sets are identical, then supposedly the expectations based on them will also have to be identical.

These are not trivial assumptions. Unfortunately, when equilibrium model builders think there may be a problem with expectation formation, it is usually attributed entirely to inadequacies of the information set rather than to the reliability of the individual's presumed method of learning in the model.[12] That is, whenever it is possible to form two different sets of expectations given

---

12. Depending on how the model builder characterizes the learning method, which may include the model used for estimation if the learning or forming method used involves econometric estimation.

any one information set, it is presumed that such a possibility is evidence of the insufficiency of the information set rather than of any inadequacy of the presumed learning method.[13] The learning method is presumed to be infallible and unambiguous whenever the number of observations is deemed to be adequate.[14] Thus in this case, it does not matter who perceives the information set, the conclusions reached are the same. The supposed inevitability of the model builder and the individual decision-maker reaching the same conclusion regarding the individual's optimum was the primary basis for the 1970s macroeconomic model builders' use of the so-called Rational Expectations Hypothesis.

### 9.1.2. The Rational Expectations Hypothesis in macroeconomic equilibrium models

Today almost all macroeconomic equilibrium model builders usually recognize the need to deal with expectations especially when addressing the disequilibrium process of attaining an equilibrium or when dealing with versions of the Arrow-Debreu model that explain choices and prices over many time periods. And today a common way to address the method used by the agents in the equilibrium model is to see them as expert econometricians when dealing with any forecasting of the future situation. But too often the model builders presume there is one best way to employ econometrics and thus every agent is using that method – as if all agents in one's model have at least a Masters degree in economics![15] Using econometric models to do forecasting has had a

---

13. If the learning method involves processing statistical data, then this would be a matter of a statistically significant difference.

14. If the adequacy requires an infinite number of observations, one might cynically suspect that these equilibrium model builders are assuming the decision makers in the model are forming expectations like one of Hitchcock's trainees discussed at the beginning of one of his 1955 TV programs. Specifically:

> You, of course, have heard the theory that if a room full of monkeys were allowed to type for a million years they would eventually reproduce all the Classics in the British Museum. This is not so. We have tried it. And, while the stories they wrote were quite good (and many of them publishable), they were not Classics – yet.
>
> This gentleman [pointing to a monkey seated at a typewriter on stage] is one of our trainees. He types nothing but gibberish. But he is not to be faulted for this. His ideas are quite good and he has a flair for dialogue – he just can't type!

15. While this might seem strange, it is exactly to what Evans and Honkapohja [2001, p. 385] were calling attention.

long problematic past.[16] And when macroeconomic model builders try to provide microfoundations for their models, too often they resort to using a representative agent[17] at the expense of the diversity found in real economies.[18] When it comes to considering learning methods, the question of diversity does not arise if the model builder thinks there is only one method of learning, since all agents would have to use that method.

Model builders who presume the model's agents use an econometrically estimated model of the economy (or of just a market) to form their expectations can easily justify such use on the ground that the information set involves imperfect data.[19] The imperfections can be caused by random variation in the accuracy of the observations that provide that data. And when dealing with such stochastic data, econometrics is seen to be the best tool for learning about the economy. But, where do the agents get the model of the economy they use to estimate in order to learn their expectations rationally? Does every agent use the same model?

### 9.1.3. The Rational Expectations Hypothesis and various ideas about learning in macroeconomics

The one thing that was assured when macroeconomic equilibrium model builders began including the Rational Expectations Hypothesis in their models was that they would be recognizing a role for learning in their models for achieving a Rational Expectations Equilibrium. How the agents in the macro model learn is no longer the simple notion that all agents are econometricians armed with a perfectly reliable estimation method and

16. As I explained in Boland [2014, ch. 7], despite what econometric model builders think, few business forecasters would ever use econometrics. Even some economists have found difficulties when they tested econometric forecasting methods – see Spyros Makridakis and Michele Hibon [1979, 2000]

17. Using a representative agent means representing all the individuals in the macroeconomy with a singular individual. The usual reason for this is to provide microfoundations for any macroeconomic conclusions. Providing microfoundations is one way to address the need to satisfy the requirement of methodological individualism.

18. For those interested in the use of the representative agent in macroeconomic equilibrium models, see Kirman [1992] and Boland [2014, ch. 1] as well as Hoover's 2012 book on microfoundations.

19. For readers unfamiliar with graduate economics and as noted in Chapter 5, when economists talk about econometric estimation, they are talking about constructing a model of the economy and then, by means of a form of reverse engineering, using available data to calculate what the parameters of that model would have to be to have 'generated' the observed data.

thus they all reach the same conclusions concerning expectations (allowing for the usual econometrics and thus stochastic variation). As was explained in Chapter 6, today we find many other ideas being included in the models to deal with the need to recognize the models' agents' learning or knowledge. Except for a few models that assume the agents are theoretical economists armed with true theories about the future, almost all are employing the 'inductive approaches'. About this, Michael Woodford [2013, pp. 304–6] observes:

> Within the category of inductive approaches, one may distinguish two important sub-categories. A first class of approaches specifies the beliefs that should be regarded as reasonable by predicting the patterns that people should be able to recognize in the data on the basis of the rationality of the procedure used to look for such patterns. A different class of approaches specifies a degree of correspondence between subjective and model-implied probabilities that should be expected, without explicitly modeling the process of inference through which such beliefs are formed. The first class of approaches [involves] models of econometric learning ... and the second class [involves] theories of partially or approximately correct beliefs.

In general, these approaches involve some form of 'inductive learning' or induction-like learning which includes econometric learning, Bayesian learning or adaptive learning. I call them induction-like simply because they are explaining the agents' knowledge by basing the notion of knowledge on the quantity-based conception of knowledge I also discussed in Chapter 6.

It is important to keep in mind that virtually all of the models being built that deal with some form of inductive or induction-like learning today are theoretical equilibrium models for which the main concern is theoretically assuring equilibrium stability.[20] While it might be very unrealistic to assume all agents in a macro model should be about as smart as (good) economists and thus to model the agents as if they were economic theorists, it does not seem to matter to these model builders. Obviously, assuming that their agents employ impossible induction or induction-like learning methods also does not seem to bother them. Instead, it seems, some just worry about the stochastic nature of observations used in the learning process as I will discuss next.

---

20. For a discussion of the distinction between theoretical and empirical models, see Part I of Boland [2014].

## 9.2. STOCHASTICISM AND MACROECONOMIC EQUILIBRIUM MODELS

The characterization of expectations as being formed rationally can be interpreted in two opposing ways.[21] One way would emphasize the economists' meaning of rational whereby the formation process is something like maximizing in the narrow sense that if agents in the model form rational expectations, those expectations would be unique (like a consumer's choice is unique when utility is maximized). The uniqueness is the basis for explaining choices. The other way is considering agents to be either econometricians or Bayesian learners dealing with stochastic information sets – as suggested by Xavier Vives [1993, pp. 329–30]:

> The literature on learning and convergence to rational expectations splits naturally into rational, and 'irrational' or boundedly rational learning models. In the former, agents have a correctly specified model of the economy and of the learning process and update their beliefs and take actions accordingly. In the latter, agents maintain incorrect hypotheses on the face of the evolution of the economy and use 'reasonable' updating procedures, like least-squares estimation, for example. It has been found that rational learning tends to yield convergence to the rational expectations equilibrium (at least in terms of convergence of beliefs) while with bounded learning convergence is obtained typically only for certain regions of the parameter space. In any case there is a paucity of results on the rate of convergence to rational expectations equilibria (which may have implications for econometric work).

In this latter way, expectations involve probability distributions of prices rather than unique sets or vectors of prices.

### 9.2.1. Rational expectations vs. bounded rationality

Thomas Sargent in his 1993 book has advocated the probability perspective on forming expectations. His reason invokes something like Herbert Simon's bounded rationality[22] in the sense that agents not only are dealing with imperfect or stochastic data, they are also limited in their ability to be the so-called rational decision maker of the Economics 101 textbooks. As a result, the outcome of an individual forming expectations of a future price means that not

---

21. See Sargent [1993].
22. See Chapter 6, fn 4.

only will the expectations be seen amounting to range estimates rather than exact values, but no two people will be necessarily forming the same expectations. In other words, invoking bounded rationality assures at least some diversity in the macroeconomic equilibrium models that attempt to address the need to explain the disequilibrium process leading to an eventual stable equilibrium attainment.

Those macroeconomic model builders who try to include some form of microfoundations to meet the requirements of methodological individualism have to be sure this inclusion does not lead to a reduction of diversity. Diversity surely is lost when macroeconomic model builders resort to using a singular representative agent to provide what looks like microfoundations merely because microeconomic theory can be used to explain the behaviour of that singular agent.[23]

### 9.2.2. Rational expectations and Bayesian learning

Over the last fifty or more years, building stochastic empirical macroeconomic equilibrium models has been very popular – particularly so since most graduate students and their professors find building stochastic macroeconomic models to be a much more convenient and a possibly less demanding activity than trying to invent theoretical models to explain how individuals make decisions based on supposedly inadequate information. The popularity is easy to explain. The students make a rational choice to spend their time processing available empirical data which they feed into computers to reduce the data to commonly understood parametric statistics. The explanation is that the students form expectations about the benefits and the probability of success of such model building and compare them to the benefits and probabilities that are expected of theoretical speculations. Obviously, the retreat to building stochastic macroeconomic equilibrium models – or very often today to building DSGE models[24] – is a defeatist position with respect to the many major questions posed by the microeconomic equilibrium model builders concerned with how we are to explain the process by which individuals make decisions in a disequilibrium setting.

Can such defeatism be avoided? There is one group of economic model builders who think they have a way.[25] Some of them would say that the

---

23. See Alan Kirman's 1992 extended criticism of the use of representative agents.
24. I briefly discussed these in Section P.3 of the Prologue and will discuss them more in Chapter 10.
25. For one example, see Bullard and Suda [2011].

individual learns by forming expectations about the likelihood of certain future states of the world and then that individual sets about updating those opinions in the light of new evidence. Many of them are assumed to use the Bayesian learning theory that I briefly discussed in Chapter 6.[26] This so-called theory of learning is just a sophisticated version of the old inductive learning theory that has been causing the problem in the first place. However, this Bayesian version is thought by some model builders to be an improvement because it relaxes the view that 'facts speak for themselves' by admitting that the learning process begins with something more than just facts. This modified inductive learning process begins with the individual's opinion or expectation after which the individual is supposed to learn from newly collected data by systematically updating or revising the original expectation. The source of the original opinion does not matter – it may even be a priori – but it is a necessary starting point.

Obviously, the nature of the original opinion is nevertheless the key aspect of the Bayesian learning process. In the simple form of Bayesian learning, an opinion is nothing more than an estimate of the probability of occurrence of a future event or of a future value of a specific variable. Since the initial opinion or expectation does not have to be based on facts, the opinion is said to be a 'subjective probability'.[27] Since these rational expectations are represented as probability estimates and learning is defined as the process of updating one's estimates, the analysis of learning is usually performed using an appropriate mathematical tool called Bayes' theorem. The detailed mathematical nature of Bayes' theorem will be of no interest here since the theorem is only invoked to overcome the inadequacies of the purely inductive learning process – that is, to overcome the view that *only* collected observations (data) matter.

One could take a more complex view of Bayesian learning by having the individual form a subjective estimate of the true objective probability *distribution* for any variable in question. Learning, in this complex view, occurs as the subjective distribution converges to the objective distribution much as Stigler [1961] saw the issue. This view, unfortunately, simply accepts an incomplete explanation of any variable unless one accepts Stigler's limited cost/benefit economics approach. It also opens the concept of Bayesian learning to an infinite regress since the individual can have an opinion of the probability distribution of the probability distribution of the probability distribution . . . ad infinitum.

---

26. See also Anwar and Loughran [2011], Nachbar [2001] and Binmore [2007].
27. See Poirier [1988] or Lancaster [2004].

### 9.2.3. Rational expectations and econometric learning

The assumption of Bayesian learning in many ways begs more questions than it answers. And worse, it is intellectually no more satisfying than the simple minded theory that one's knowledge must be based on inductive learning such that new facts alone always constitute learning even though there is no guarantee that one has 'learned' the true expectations or true knowledge.[28] But, of course, using questionable mathematical tools like Bayesian learning will not be of concern to members of the newer formalist model-building culture (recognized by Weintraub and discussed in Chapter 5) – particularly, to those who might be interested only in building formal stochastic equilibrium models. After all, as I have been noting, for too many members of that newer model building culture, questions of realism are secondary.

As I said in Section 9.1, today's formalist equilibrium model building culture includes building theoretical macroeconomic models that deal with learning by assuming all agents are expert econometricians. In these models agents learn by first building a model of the economy and then by using available data to estimate the values of the model's parameters. They then proceed to use their model to forecast future prices. Much like Bayesian learning, econometric learning involves updating a prior opinion (in this case, a conjectured model) by collecting more data. Unlike with the Bayesian learning (that just involves updating a prior opinion), with econometric learning the opinion is a more elaborate and structured opinion. And for this reason, econometric learning is also constrained by the limits of using econometrics to do forecasting.

The main problem with using econometrics to do forecasting is that the parameters of any econometric model are usually assumed to be ergodic (i.e., they are assumed to not change) – not only over the time period for which data was collected to calculate their value, but also extending over any future period of time for which the econometric model is used to generate predictions of future prices. But to assume such ergodicity begs too many questions. Of course, one could limit one's predictions to just a very short future time period ahead but not much would be learned that way.

The idea that all firms in one's equilibrium model employ expert econometricians is a rather heroic assumption at best and at worse is simply false, given what I learned in my investigation when writing Chapter 7 of Boland [2014] about forecasting using econometric models. What I learned was simply that econometrics-based forecasting is usually rejected as a means of doing

---

28. I should point out here that while there may be considerable reasons to reject subjective probabilities and Bayesian learning, one can still accept that *some* agents in one's equilibrium model can be assumed to think they learn this way. Any criticism discussed here is just that there is no reason to assume *everyone* in the model makes decisions this way.

business forecasting.[29] But of course, as critics of the old culture will complain, whenever the newer culture are building their formal equilibrium models, realism is less important than mathematical elegance.

## 9.3. INSTRUMENTALISM AND THE USE OF STOCHASTICISM IN EQUILIBRIUM MODELS

Perhaps some equilibrium model builders will not like my use of the idea of *true* expectations or *true* knowledge or even realism as a basis for rejecting techniques that characterize learning using the Bayesian approach or the econometrics learning approach. These model builders may say that such a basis for criticism is misleading since nobody could ever have true expectations or true knowledge if they are learning inductively or in a Bayesian manner or anytime it is assumed that bounded rationality is a useful basis for rational expectations. They may say, in effect, that true knowledge or the formation of true expectations would require a virtual infinity of information and that, as Stigler might say, would at best cost far too much. A critic of these model builders may counter that economic model builders must explain then why an individual chooses to have economically optimal but imperfect knowledge and thus form imperfect expectations. However, some model builders might even go so far as to argue that for practical purposes true expectations or true knowledge is not even needed.

For this last group of critics of my rejection, particularly economists who are followers of Milton Friedman's pronouncements about methodology,[30] equilibrium models and theories have no other purpose than to serve as instruments for forming economic policy or for forming predictions of the consequence of policies. Practical success is all that matters. Following Friedman's instrumentalism is an all too easy way of avoiding addressing complex theoretical problems, such as those posed by all the considerations of disequilibrium economics or of stable equilibrium processes. Obviously, the time required to form perfect expectations is impossibly long, but the time required to benefit from a Bayesian or econometric learning process may also exceed that available to make day-to-day practical decisions. It might even be argued that the benefits of incorporating all the sophisticated disequilibrium concepts (such as when using the Rational Expectations Hypothesis) into the ordinary neoclassical equilibrium model may be too small when it comes to questions of improving the policy-maker's estimates and recommendations.

---

29. And again, as I noted in note 16 above, more about model-based forecasting can be found in Boland [2014, ch. 7].

30. See his famous 1953 book of essays. Friedman's version of instrumentalism can be seen to exist at least as far back as the early eighteenth century when Bishop Berkeley objected to anyone considering Newton's physics to be true rather than just an instrument for predicting the rotation of the planets.

Obviously it is just as easy to say that expectations and knowledge do not have to be true for even the individual decision-maker if one thinks that economic theory is only an instrument for forming policies or making predictions. Adherents to instrumentalism reject the complex theoretical questions that are posed by the need for the disequilibrium foundations of any equilibrium model used in policy economics.[31] For many equilibrium model builders such a bold rejection is considered a matter of bad taste. In this light many equilibrium models builders may explicitly claim they do not believe in instrumentalism – but somehow actions speak louder than words. There were numerous articles published[32] claiming that instrumentalism is not the methodological view that economists advocate. Nevertheless, it is difficult to distinguish instrumentalism from the defeatism implied by stochastic macroeconomic model building.

Whether the stochasticism advocated is in the form of elaborate econometric equilibrium models consisting of simultaneous equations where each equation includes an error term thereby admitting that the equation does not have to be exactly true, or it is in the form of the bounded-rationality-based stochastic decision processes in which learning is always thought to be imperfect, it makes no difference. The justification for stochasticism is ultimately the same as the argument given for instrumentalism. For some, the reason given for advocating the development of stochastic macroeconomic equilibrium models might be that an accurate theory of the individual would be computationally too complex.[33] The usual reason for opting for instrumentalism is that a *true* short-run theory of individual behaviour (or of macroeconomic behaviour) would be too complex for practical purposes. What is suppressed, both in stochastic macroeconomic equilibrium models and in instrumentalist microeconomic equilibrium models, is just the very methodological individualism that has motivated almost all neoclassical economic theories and models for very many decades. Stochasticism and instrumentalism, as well as macroeconomics in general, are all ways to avoid the difficult questions that have been raised by those economists interested in understanding the disequilibrium foundations of economic equilibrium models or those interested in building a methodological-individualist view of the economy that does not depend on the narrow confines of neoclassical long-run or general equilibrium theory and models.

31. For a more general discussion of instrumentalism, see Boland [1997, Part 1] and [2003, ch. 5].

32. All of them were criticizing or responding to my widely cited 1979 article in which I was criticizing Friedman's instrumentalist methodology as well as criticizing all the faulty previous attempts to criticize Friedman's famous 1953 article. See Caldwell [1980], Hirsch and de Marchi [1984] and Hoover [1984] – see also Wible [1982].

33. It is often for this reason, as I discussed in Section 9.2 above, that bounded rationality (usually the one attributed to Herbert Simon) is suggested as an alternative way to characterize behavioural optimization and learning.

# CHAPTER 10

❧

# Equilibrium models intended
# to overcome limits

In this chapter I return to looking at the critical issues surrounding formal Walrasian general equilibrium models upon which many macroeconomic models are built to assure a minimum of methodological individualism. Of the two cultures I discussed in Chapter 5 concerning various concepts of an economic equilibrium – particularly that are used in general equilibrium models – the older culture sees an equilibrium state as being about a property of the real world economy. The newer culture – the one I was trained in during my PhD education – is concerned with building formal theoretical equilibrium models and, for this newer culture, an equilibrium is only a property of a model designed to explain or represent some conceivable aspect of the real world economy.

Let us begin by considering some of what can be found in *The New Palgrave: A Dictionary of Economics*. Lionel McKenzie, who is one of the first general equilibrium model builders to provide a proof of the existence, provided an entry on general equilibrium. In it he is somewhat critical as he explains [1987, p. 510]:

> [G]eneral equilibrium theory is a partial theory of economic affairs with a special set of *ceteris paribus* assumptions. The variables which are left free are chosen because they lend themselves to a particularly elegant theory in terms of consumer demand under budget constraints and producer supplies with profit conditions where these constraints and conditions are established by prices equating demand and supply.
>
> The convexity assumptions which have appeared in general equilibrium models from the time of Walras are often not good approximations of reality

though they are depended on for many of theorems of the subject, such as the theorems on existence and Pareto optimality.

Finally, the assumption that the market participants take prices as independent of their actions fails to describe many markets, and describes very few exactly.

The Arrow-Debreu version of the general equilibrium model was discussed in another somewhat critical entry, this one by John Geanakoplos [1987, pp. 116–19] who says:

> The elaboration of the nature of the primitive concepts of commodity and rational choice, developed as the basis of the theory of market equilibrium, prepared the way for the methodological principles of neoclassical economics (rational choice and equilibrium) to be applied to questions far beyond those of the market. . . . Price is the final primitive concept in the Arrow-Debreu model. Like commodity it is quantifiable and directly measurable. . . . Notice that in general equilibrium each agent need only concern himself with his own goals (preferences or profits) and the prices. The implicit assumption that every agent 'knows' all the prices is highly non-trivial. It means that at each date each agent is capable of forecasting perfectly all future prices until the end of time. It is in this sense that the Arrow-Debreu model depends on 'rational expectations'.

The primary criticism of recent general equilibrium models is the same as that which Hayek identified in his 1937 article and Richardson explained in his 1959 article. Both said that if you are going to use a state of equilibrium in your explanation, you need to explain how the participants in the equilibrium acquired the knowledge necessary to assure the equilibrium's attainment. Failure to heed this requirement continues to be a problem for equilibrium model builders whether the model is for the whole economy, for some limited game theory situation, or just a small model for an individual decision maker.

Obviously, if one's equilibrium model involves investment decisions, the model's decision makers need to form some expectations (rational or otherwise) concerning future outcomes for their investments decisions. Critics may see expectation formation as a matter of using the information available to the decision makers as I discussed in Chapter 9 or they may just see it as a matter of how they conjecture those expectations. Either way, since expectation formation is about the inherently unknown and unknowable future, the critics will claim there is still the matter of dealing with uncertainty.

In this regard, some critics claim that whenever a formal general equilibrium model presumes the economy being represented involves only perfect competition, this presumption leads to building a model where, for its equilibrium state,[1] every firm at its point of production is producing where its

---

1. Note that textbooks usually fail to point out that in terms of consequences, a Walrasian general equilibrium model is logically equivalent to a model of a Marshallian

production function is linear and homogenous or, in other words, every firm's production function exhibits constant returns to scale at that point.[2] These critics claim that such a model is unrealistic since the *real world* can more often exhibit increasing returns to scale simply because, in many cases, competition is just not usually perfect.

## 10.1. THE ALLEGED LIMITS OF GENERAL EQUILIBRIUM MODELS

In the Prologue I mentioned DSGE models and their popularity. And as Hashem Pesaran and Ron Smith observed 'Academic macroeconomics and the research departments of central banks have come to be dominated by dynamic stochastic general equilibrium (DSGE) models based on microfoundations of representative agents, who solve intertemporal optimization problems under rational expectations' [2011, p. 5]. Since the North Atlantic financial crisis of 2007 and 2008 macroeconomic equilibrium models have been subject to intense criticism and some critics simply complaining that they are irrelevant to the real world. About this, John Driffill [2011, p. 1] critically adds that the microfoundations provided by the DSGE model

> include representative households, . . . imperfect competition, rational forward-looking agents who perfectly understand their environment and have rational expectations, and good if not perfect financial markets. Asset prices are determined by fundamentals: there is no room for bubbles here. There is no lending or borrowing, since all households are the same. There are no problems connected with enforcement of contracts, default or bankruptcy. Most DSGE models have no room for the financial sector, and in some money plays little role.

Before evaluating the ability of DSGE models to overcome the limits identified by both the older culture that generally objected to the realism of any general equilibrium model and the newer culture of formalist general equilibrium model builders that worry about whether their model is logically adequate for their model's intended purpose, I need to list the alleged shortcomings. Apart from the question of realism that concerns those of the older culture, most complaints concern the ingredients too often missing in general equilibrium

economy in long-run equilibrium. And thus, as I discussed in Chapter 8, note 12, at that point in the equivalent long-run equilibrium all firms are maximizing profit but that profit is zero, which means the firm is producing at the output level at which average cost is minimum.

2. Namely, producing where output is proportional to input in the close neighbourhood of that point. This aspect of a long-run equilibrium was discussed in Chapter 8 and specifically explained in note 12 of that chapter.

models. Some of the shortcomings were discussed in Chapter 5. The main issues of interest concern the shortcomings of the usual Walrasian general equilibrium model. As I noted in Chapter 6, those of us who were trained in mathematical model building in the 1960s were led to believe we were talking about Walras' version of general equilibrium[3] but, as Weintraub [1985a] noted, we were really talking about a simplified version that Gustav Cassel [1918/23] presented in his quasi-textbook, *Theory of Social Economy*.[4] It is mostly Cassel's version that is being talked about when economists talk about Walrasian general equilibrium – particularly the linear-equation version.[5]

The main problem with the Cassel version of Walrasian general equilibrium is that it concerns only a so-called stationary equilibrium, as it does not address the issue of dynamics or time.[6] To a limited extent, Hicks did address these issues. He did this by distinguishing markets for any specific good by when the market operates, such that those at different points of time were considered markets for different goods. Think of this as putting a time index on each good: $X_1$, $X_2$, $X_3$, ... $X_t$: the index numbers can refer, for example, to a sequence of years. He also recognized that if the index refers to different points in time, it is possible that decisions being made in year 1 might affect markets in future years and thus future prices.[7] Given such a recognition for a role of time in a general equilibrium model, Hicks said the model's decision makers need to form expectations.

Not long after the second edition of Hicks'1939 book appeared in 1946, several general equilibrium model builders focused on providing a proof of the existence of his type of general equilibrium. The 1954 articles by McKenzie and by Arrow and Debreu provided such proofs. Today, much of the discussion involves the Arrow-Debreu model of general equilibrium[8] that addresses market equilibria for a finite set of commodities and a finite set of future

---

3. And worse, many of us thought that Walrasian economics was the center of attention in economic model building throughout the first half of the twentieth century when actually it began mostly with the 1939 book by Hicks who set about developing a version of general equilibrium. Moreover, as Malcolm Rutherford [2001] has explained, until the 1950s, old institutionalism (which is concerned with observable data about economic institutions rather than building theoretical models) dominated mainstream economics departments in North America.

4. Cassel's book was published in German in 1918 and first appeared in English in 1923.

5. For example, see Dorfman, Samuelson and Solow [1958, ch. 13].

6. See De Vroey [2002].

7. This was also the issue in Hayek's 1933 lecture that I discussed in the Prologue. In particular, a decision made in the investment market may turn out to be sub-optimal at some later market, say at year 3. If this is the case, it puts into question whether there really was an equilibrium in year 1.

8. To give full credit, some writers refer to this model as the Arrow-Debreu-McKenzie model. For full discussion of this history of this model, see Düppe and Weintraub [2014].

markets. And like Hicks, their model defines commodities by time and location and thus expands the general equilibrium well beyond the size of the Cassel-Walrasian general equilibrium model. And as John Geanakoplos points out in the 1987 edition of the *New Palgrave Dictionary*,[9] the Arrow-Debreu model also needs to recognize the expectations of the decision makers. This is primarily because all decisions are made on the first day of a finite number of time periods being considered in the model. As Geanakoplos also points out [1987, p. 120]:

> Observe finally, that although the commodities may include physical goods dated over many time periods, there is only one budget constraint in an Arrow-Debreu equilibrium. The income that could be obtained from the sale of an endowed commodity, dated from the last period, is available already in the first period.

I turn now to how time and dynamics are actually treated in general equilibrium models, since it has been of concern for many years.

### 10.1.1 Dynamics vs. time

> The Hatter was the first to break the silence. 'What day of the month is it?' he said, turning to Alice: he had taken his watch out of his pocket, and was looking at it uneasily, shaking it every now and then, and holding it to his ear.
> Alice considered a little, and then said 'The fourth'.
> 'Two days wrong!' sighed the Hatter. 'I told you butter wouldn't suit the works!' he added looking angrily at the March Hare.
> 'It was the BEST butter', the March Hare meekly replied.
> Lewis Carroll [1871/85, pp. 57–58]

While the critics from the older culture often complain that the Walrasian general equilibrium model is timeless, admirers of the Arrow-Debreu model can easily point out that this is not true since it involves very many time periods as well as expectations spanning many time periods. But are the admirers correct? It would seem that Arrow seems to think they are as he has noted [1974, pp. 265 and 268]:

> [W]orking independently and in ignorance of each other's activities, Debreu and I both started applying Kakutani's fixed point theorem to the problem of

9. While his article also appears in the 2008 second edition of the *New Palgrave Dictionary of Economics*, there do not seem to be many differences if any between his entries in the two editions, so I will only be referring to the 1987 edition here because with it I can provide specific page numbers.

existence. . . . What we are led to is considering the same physical commodity in different states of the world as economically different commodities. The procedure is exactly analogous to Hick's analysis of present and future goods (1939); the same physical commodity at different points of time define different commodities.

Surely, though, just putting a time index on a commodity is logically no different than putting on an index that represents different locations of the markets for that commodity. That is, whenever one is discussing a hamburger sold in today's market as a different hamburger than one that will be sold in tomorrow's market it is no different logically than discussing a hamburger sold today in London's hamburger market as being different than a hamburger being sold today in Rome's market. So, it does not matter that the Arrow-Debreu model might include markets located in different places and different periods. It is just an index. And as such, it is not a model *in* time – it is a model in which 'time' is just an interpretation of a mathematical index. Hicks [1976, pp. 141–42] – referring to what he did in his 1939 book – discussed what he thought should be considered for a model to truly involve dynamics:

> While Keynes had relegated the whole theory of production and prices to equilibrium economics, I tried to keep production *in* time, just leaving *prices* to be determined in an equilibrium manner. I wanted, that is, to go further than Keynes, keeping closer to [Erik] Lindahl. But I could only do so by an artificial device, my 'week,' which was such that all prices could be fixed up in what would now be called a 'neo-Walrasian' or 'neo-classical' manner, on the 'Monday'; then, on the basis of these predetermined prices, production *in* time could proceed. It was quite an interesting exercise; it did bring out some points – even some practically important points – fairly well; but I have become abundantly conscious how artificial it was. Much too much had to happen on that 'Monday'! And, even if that was overlooked (as it should not have been overlooked) I was really at a loss how to deal with the further problem of how to string my 'weeks' and my 'Mondays' together.

For a model to be *in* time, endogenous changes within the model intrinsically must not be reversible. Since time in the Arrow-Debreu model is merely an interpretation of a commodity's index – an index that can also be interpreted as a space-location index for which one can move back and forth between such locations – the time in the Arrow-Debreu model does not satisfy what Hicks was demanding.[10] Admittedly, not

10. In Boland [1982 and 2003] I offered a way to satisfy Hicks. My notion was to recognize knowledge as a variable that is irreversible (usually, one cannot unlearn) and thus recognizing knowledge in one's model would be a way to satisfy Hicks.

everyone demands what Hicks was demanding but nevertheless he was not alone.[11]

Despite the above quotation from Arrow's 1974 article, some model builders might think the Arrow-Debreu model includes time, but there is no dynamics that explains how prices are determined and thus dynamics that would have satisfied what Arrow demanded in his 1959 article.

### 10.1.2  A list of other short-comings of the Arrow-Debreu general equilibrium

Money has been seen as a problem for general equilibrium models for a long time. As Hahn [1965, pp. 126 and 134] observed:

> Recent work on the existence of an equilibrium has been concerned with a world without money while all work in monetary theory has ignored the 'existence' question.
>
> In a society where contracts are made in terms of money and recontract is not possible an equilibrium solution may not exist. Of course in the 'long run' all contracts are escapable but that is another story.
>
> The point just made … focuses attention on what is probably one of the most important features of a monetary economy: namely that contracts are made in terms of money.

As Douglas Gale [2008] observed in his entry in the second edition of the *New Palgrave Dictionary* on money and general equilibrium,

> Hahn [1965] pointed out [a] problem in the theory of monetary equilibrium, viewed from the Walrasian perspective. The problem was the lack of a proof that money has positive value in equilibrium. Hahn observed that the uses of money that might be expected to give rise to a positive demand for money all require money to have positive value in exchange. If the value of money were zero, the economy would be identical to a barter economy. Under the usual assumptions … such a non-monetary economy would possess an equilibrium, but it would not be a monetary equilibrium, because money would have no role in exchange. . . .

---

11. Robinson [1974] also complained about neoclassical models not recognizing that history is irreversible.

As Joseph Ostroy wrote in the first edition of *The New Palgrave* [1987, p. 515],

... We shall argue that the incorporation of monetary exchange tests the limits of general equilibrium theory, exposing its implicitly centralized conception of trade and calling for more decentralized models of exchange.

That comment is just as true today as it was then, and remains the great challenge for economists who want to develop more satisfactory models of the process of monetary exchange at the level of the economy as a whole.

In the Geanakoplos entry on the Arrow-Debreu model of general equilibrium, he conveniently devoted his last section to a list of things that the Arrow-Debreu model 'does not explain'. While the Arrow-Debreu model involves markets for commodities, it does not include 'trade in shares of firms'. About which Geanakoplos notes [1987, p. 121]:

in Arrow-Debreu equilibrium, the hypothesis that all prices will remain the same, no matter how an individual firm changes its production plan, guarantees that firm owners unanimously agree on the firm objective, to maximize profit. If there were a market for firm shares, there would not be any trade anyway, since ownership of the firm and the income necessary to purchase it would be perfect substitutes.

And he also notes, 'Bankruptcy is not allowed in an Arrow-Debreu equilibrium. That follows from the fact that all agents must meet their budget constraints' [pp. 121–22]. And most important, Geanakoplos notes, 'Money does not appear in the Arrow-Debreu model. Of course, all of the reasons for its real life existence: transactions demand, precautionary demand, store of value, unit of account, etc. are already taken care of in the Arrow-Debreu model' [p. 122]. He goes on to explain,

There is no point in making the role of money explicit in the Arrow-Debreu model, since it has no effect on the real allocations. However, if one considers the same model with incomplete asset markets, the presence of explicitly financial securities can be of great significance to the real allocations.

But for money to play a significant role beyond serving as a numeraire, time would have to matter as Hicks advocated.[12] And separately Shackle [1972] and Davidson [1972, 1977] explained what it would take to give money a role beyond just serving as a numeraire. They both argued that if equilibrium models such as the Arrow-Debreu model were to address real clock and calendar time there

---

12. For an explanation of the role for money, see Boland [2014, pp. 54–55].

would be a need for money to play a real role. For Shackle and Davidson, money can always play a significant role as a means of enforcing contracts, particularly market contracts for transactions that are unlike barter transactions. The non-barter transactions they were discussing are those in which one places an order at one point in time for some amount of a commodity but does not expect to receive that amount until some later point in time. Usually, a contract would be drawn up that might specify a monetary penalty for failure to provide the specified amount on time. This, I think, means that an Arrow-Debreu model could give money a significant role only if it included contracts and the possibility of failure. But, if, as Geanakoplos seems to suggest, rational expectations in the Arrow-Debreu model where the presence of multi-period decisions implies perfect knowledge regarding all future prices and transactions, why would we need contracts? Obviously, one might conjecture that this is the real reason why there is no significant role for money in an Arrow-Debreu model.

Geanakoplos sees this problem concerning a role for money differently. For him, the complaint is that the Arrow-Debreu model does not consider incomplete asset markets.[13] In this regard, like many other commentators, he notes the one big problem is that 'all trade takes place at the beginning of time' in the Arrow-Debreu model. Moreover, 'the Arrow–Debreu proofs of the existence and Pareto optimality of equilibrium do not apply to such an incomplete markets economy' [Geanakoplos 1987, p. 122].[14] At best it means the Arrow-Debreu model is limited in its view of a market economy.

The major point about the Cassel-Walras and Arrow-Debreu models is their focus being on only the maximizing behaviour of individuals and as Geanakoplos notes, their 'celebrated conclusions about the potential efficacy of unencumbered markets'. He goes on [p. 122]:

> But still more telling is the fact that the assumption of a finite number of commodities (and hence of dates) forces upon the model the interpretation of the economic process as a one-way activity of converting given primary resources into final consumption goods. If there is universal agreement about when the world will end, there can be no question about the reproduction of the capital stock. In equilibrium it will be run down to zero. Similarly when the world has a definite beginning, so that the first market transaction takes place after the ownership of all resources and techniques of production, and the preferences of all individuals have been determined, one cannot study the evolution of the social norms of consumption in terms of the historical development of the relations of production.

13. In general terms an incomplete model is one for which not all of the necessary conditions for a market to exist are satisfied. The simplest case of an incomplete market would perhaps be one with an excess demand for some sort of public good such as a road or bridge.
14. He credits Oliver Hart [1975] for first pointing this out.

A common complaint about general equilibrium models (as I noted in Chapter 5) is their failure to address questions about the role of information. Geanakoplos raises this in a slightly different form. He complains, 'There is no place in the Arrow-Debreu model for asymmetric information' [p. 122]; and

> [I]n the definition of equilibrium no agent takes into account what other agents know, for example about the state of nature. Thus it is quite possible in an Arrow-Debreu equilibrium for some ignorant agents to exchange valuable commodities for commodities indexed by states that other agents know will not occur. This problem received enormous attention in the finance literature, and some claim (see [Sanford] Grossman 1981) that it has been solved by extending the Arrow-Debreu definition of equilibrium to a 'rational expectations equilibrium' . . . But this definition is itself suspect; in particular, it may not be implementable.

Geanakoplos finished his list with two major missing items in Arrow-Debreu models. The first of these is what I noted at the end of the Section 10.1.1 – namely, that the Arrow-Debreu model does not include any explanation for how the model's agents get to the equilibrium (viz., the model is missing what Arrow discussed in his 1959 article). This is always a problem for the Arrow-Debreu model because transactions do not take place until the equilibrium is reached presumably after an ongoing bidding process of some type. About this Geanakoplos says, 'This illustrates a grave shortcoming of any equilibrium theory, namely that it cannot begin to specify outcomes out of equilibrium' [p. 123]. This means that the applicability of equilibrium analysis is very limited whenever it is applied to economic crises such as what occurred in the 1930s and to what occurred in the more recent Great Recession of 2007–8.

The other missing item concerns attempts to apply game theory to general equilibrium. While one can understand why formal model builders might want to use the mathematical techniques provided by using game theory, all of the informational problems cannot be avoided with game theory characterizations such as the Nash equilibrium,[15] and as Geanakoplos notes, 'the informational requirements of Nash equilibrium are at least twice that of Arrow-Debreu competitive equilibrium' [p. 123].

## 10.2. THE CURRENT ATTEMPTS TO OVERCOME THE LIMITS OF GENERAL EQUILIBRIUM MODELS

While Walrasian general equilibrium models were usually thought to be microeconomic models, microeconomic equilibrium model builders seem to have

---

15. For a broader discussion about problems with applications of game theory, see Boland [2014, ch. 4].

lost any major interest in analyzing the Arrow-Debreu model and its proof of equilibrium existence. Where one does find interest in that model it is among macroeconomic equilibrium model builders since, after all, a general equilibrium is also about the whole economy. But the primary reason is that the Arrow-Debreu model is seen as a means of providing microfoundations for macroeconomic equilibrium models. And most of the work on fixing the various shortcomings listed by Geanakoplos [1987] of the Arrow-Debreu model is being done in terms of addressing macroeconomic issues of its application. Today, the so-called Dynamic Stochastic General Equilibrium (DSGE) model is the most popular model to use for basic microfoundations and, in this regard, it is just the latest version of the Cassel-Walras equilibrium model.[16] And of course, it primarily relies on the Arrow-Debreu version of that model suitably augmented to characterize the dynamics of a general equilibrium analysis and thereby applied to specific macroeconomic questions such as those involved in business cycle theory. But to use it, most of its shortcomings need to be dealt with.[17]

### 10.2.1 DSGE models to the rescue?

Many of the alleged shortcomings listed by Geanakoplos are dealt with by adding assumptions and variables to the Arrow-Debreu general equilibrium model. Some find that assuming rational expectations in the Arrow-Debreu model allows treating knowledge of future prices as being perfect knowledge.[18] But it is easy to see that the assumption of rational expectations actually requires recognition that such expectations can be statistically less than perfect. After all, the usual econometric estimations of future events involve 'error terms' in the estimated equations. But this is not the usual reason why the DSGE model involves stochasticism. Today, the usual DSGE model is stochastic, unlike the Arrow-Debreu version, because the DSGE models presume demanders are maximizing utility after forming rational expectations for all future markets – thus, today's stochastic version usually presumes demanders maximize *expected* utility.[19]

16. For a very informative history of the development of the DSGE model, see Pedro Duarte [2011].

17. And some general equilibrium model builders will note that some of the shortcomings are ignored, such as the results of the Sonnenschein-Debreu-Mantel theorems, as I explained in Boland [2014, ch. 2].

18. But, as discussed in Chapter 9, this treatment is based on a neoclassical interpretation of the adjective 'rational' such that, as noted by Sargent [1993], rational means maximization and thus implies a unique outcome.

19. For those readers unfamiliar with this jargon, expected utility is nothing more than one's utility multiplied by what one thinks is the probability of obtaining that utility.

The immediate predecessors to the DSGE model were models based on Real Business Cycle (RBC) theory. Most RBC model builders were followers of Robert Lucas and thus use the mathematical assumption of rational expectations using the 1961 article by Muth.[20] The assumption still involves a degree of information imperfection but only in that it assumes all decision makers have the same imperfect information as that available to the model builders. As was discussed in Chapter 9, the use of Muth's Rational Expectations Hypothesis amounts to assuming that all decision makers are expert econometricians and thus their expectations will not deviate from what practicing econometricians and their included error terms would expect of future prices and events.[21] The 'Real' in RBC models refers to the assumption that money plays a minor role (if any) in the determination of the values of real variables such as the level of employment or of real national income.[22] The RBC models were offered as alternatives to the typical Keynesian model, in which such things as a monetary change in liquidity preference[23] can have an effect on the level of employment or on the investments in productive capital to expand the size of a firm.

Today, DSGE models are of interest to both RBC model builders and so-called new Keynesian model builders.[24] To a certain extent, the DSGE model was the result of efforts to form a synthesis that would allow their different perspectives to be applied to a DSGE model by augmenting it to incorporate one or other of their perspectives. Having both perspectives using the same basic general equilibrium model as their starting point for model building is

20. About this version of a RBC model, see Long and Plosser [1983] and Dotsey and King [1988].

21. Benjamin Friedman [1979] also noted this.

22. For those unfamiliar with the jargon term 'real' here, this means that the dollar value of the national income is divided by the current consumer price index so that inflation does not distort year by year comparisons of data concerning real variables such as one hour of labour or one unit of some specific type of machine.

23. For Keynes this meant that people might hold money for reasons other than just buying commodities. Holding money for liquidity meant that some people might think they would need extra money specifically for the purpose of speculation or just precaution. Keynes also recognized circumstances where monetary policies (such as increasing the money supply) might not have an effect whenever the going interest rate is extremely low (as it was following the 'Great Recession' of 2007–8). He called this the 'liquidity trap' because, given the low interest rate and thus that any monetary gain from putting one's savings in the bank is low, any increase in the money supply would just go into people's pockets and not into the banking system that might provide the money to investors. This might also occur when everyone sees prices of durables falling and people just wait to buy because tomorrow the prices will be lower and thus they leave their money in their pockets. This goes for investors, too. Who wants to expand their production when prices are falling? I will return to discuss all this in a broader context later in Chapter 14).

24. Even though the original Keynesian model builders came from a radically different ideological perspective. If there is a difference in perspective today, it might just be that the new Keynesians allow for the government to play a more prominent role in their models.

seen to amount to a consensus having been reached – and many see this as a major positive step in the progress of economic theory and model building. Presumably, in this way many of the shortcomings of equilibrium models can thus be overcome by using a DSGE as the starting point for model building or the ideological differences might at least be lessened between the RBC and new Keynesian model builders.

Clearly the inclusion of Muth's stochastic version of rational expectations addresses the complaint that general equilibrium models lack some sort of recognition of *uncertainty* and *expectations*. And relatedly, at least *information* is being recognized. Unfortunately for some, recognizing information seems to be at the expense of diversity. Moreover, some versions of the augmented DSGE model may include increasing returns by including *imperfect competition* and thereby recognize *increasing returns* as an equilibrium outcome. And finally, since the DSGE model is used to build empirical general equilibrium models for the purpose of applying them to government policy assessments, it does not necessarily suffer from not being *operational* as Machlup complained in his 1958 article.

For convenience, many builders of a DSGE model wish to use it to recognize an equilibrium state as one being based on the behaviour of individuals and thus, as I have noted, they use it to serve as a basis for microfoundations for their macro applications – and for a few a DSGE model provides even the possibility of satisfying the requirements of methodological individualism. But many macroeconomics equilibrium model builders conveniently take the model a step further by using a representative agent to represent some or all individuals in the DSGE model. They do this with either one or two representative agents.[25] Unfortunately by doing so all of the diversity that could be recognized in a DSGE model is lost merely for what seems to be mathematical convenience.[26]

### 10.2.2 Recent unrealistic efforts in formal equilibrium model building to address limitations

In his 2008 *New Palgrave Dictionary* entry on 'general equilibrium (new developments)', Zame complains:

> For many purposes, it is not enough to know simply that competitive equilibrium exists; we would like to know how equilibrium varies when the underlying

---

25. Sometimes just one for all individuals and at other times one for all consumers and another for all firms.

26. Alan Kirman [1992] explains how using the representative agent can lead to even more problems – problems that are necessarily involved in the mathematics of such a model, such as the so-called Sonnenschein-Debreu-Mantel results I mentioned in note 17.

parameters of the model vary. Such comparative statics analysis is simplest and most convincing when equilibrium is unique and depends nicely on the underlying parameters. However, it has been known for a long time that even some very simple economies admit multiple equilibria and that conditions on the primitives of an economy that guarantee uniqueness of equilibrium must necessarily be unpleasantly strong.

However, his entry provides discussions of aspects of general equilibrium models such as 'determinancy', 'perfect competition and price-taking', 'equilibration', 'incomplete markets', 'infinitely many commodities' and 'hidden information and hidden actions'. The members of the older culture discussed in Chapter 5 will not likely be impressed with the developments that Zame discusses as they are all about various formal mathematical assumptions added to the Arrow-Debreu model. And he discusses the inclusions of assumed behaviour on the part of agents in the model that help solve some of the formal mathematical problems of providing proofs of equilibrium existence by assuming behaviour that involves pooling, lotteries, adverse selection, convexity, fixed-point theorems, etc. The realism and relevance of the many efforts Zame discusses are likely to be considered questionable by those economists interested in applying such models to real world data. But of course, such concerns are usually and perhaps rightly secondary for formal equilibrium model builders.[27]

## 10.3. THREE EMPIRICAL ALTERNATIVES TO WALRASIAN GENERAL EQUILIBRIUM MODELS

As I noted in Chapter 5, one critic specifically complained that the usual theoretical general equilibrium models fail to be 'operational'. That is, they do not involve observable data but are only formal mathematical exercises concerned with existence proofs, uniqueness proofs and proofs of stability of the equilibrium solution.[28] I noted, however, there are some model builders who might object to this complaint (about whether equilibrium models are operational) by pointing to existing 'computable general equilibrium' models and 'applied general equilibrium' models used by many governments and organizations such as the World Bank. For example, Arrow [2005, p. 13] observes,

> The applied (or computable) general equilibrium (CGE) model is one of today's standard tools of policy analysis. . . . The ability to deploy CGE models is the

27. For any readers interested in knowing about recent efforts to address the limitations itemized by Geanakoplos, Zame's *Palgrave* entry is nevertheless very informative and useful.

28. As I point out in Chapter 5 (and at length in Boland [2014]), there are two types of model: theoretical and empirical. Builders of theoretical models are not interested in applying their models to empirical data so this criticism is irrelevant.

outcome of research going back at least 130 years and involving very disparate lines of inquiry. Economic theory and the vastly improved availability of economic data have played basic roles. But other research inputs have been equally crucial: improvements in computing power and the development of algorithms for computing equilibria. The decisive step in the last direction has been the pioneering work of Herbert Scarf.

However, some critics might still disagree, saying that those models are not really general equilibrium models.

Years before Hicks introduced his version of a Walrasian general equilibrium model, Wassily Leontief created the so-called Input-Output model. This model, like Walrasian general equilibrium models, accounts for all transactions in the whole economy, but, unlike general equilibrium models, the Input-Output model is limited to the aggregate inputs and outputs of all transactions. However, while many may consider the complete Input-Output model to be an equilibrium model, it is not, because usually it is not used to determine prices – which the basic Walrasian model is designed to do.

Today two of the recognized basic empirical alternatives to Walrasian general equilibrium models are the so-called Computable General Equilibrium (CGE) model and the Applied General Equilibrium (AGE) model. These two models started out as different approaches but are now often referred to interchangeably.[29] CGE models began as basically macroeconomics models along the lines of the Leontief's models, whereas AGE models are allegedly empirical versions of Arrow-Debreu general equilibrium models.[30]

About these two types of empirical model, Benjamin Mitra-Kahn [2008, p. 53] observes that a CGE model consists of a solvable system of simultaneous macro balancing equations. Using such models, their exogenous variables can be changed outside the model in order to give differing results concerning the solved-for endogenous variables. He also notes that modes of the other type, the AGE model, are based on Arrow-Debreu general equilibria in a completely different manner. Using such a model, an equilibrium set of market-clearing prices is solved for by using the Scarf algorithm discussed in Arrow's quotation above after inputting data into all of the model's recognized sectors. This, however, can yield a large number of sets of possible equilibrium prices and thus various mathematical techniques are employed to reduce the number of sets of possible solutions. Stopping the narrowing process is basically arbitrary such that a set is chosen that is governed only the usefulness of the set's approximate solution.

Except for the matter of being empirical, whether these two alternative models overcome any of the shortcomings listed by Geanakoplos is open to

29. For example, see the above quotation from Arrow [2005, p. 13].
30. See Mitra-Kahn [2008, p. 20].

question. For governments and banks that find them useful as operational and empirical models, this may only be a matter of social engineering allowed by instrumentalist methodology[31] for which perfect realism does not matter. For some, it is often a matter of being 'close enough' for the purposes for which they are being used.

The third recognized alternative addresses the issue of realism directly even though it still involves econometric models. It differs from the first two by building equilibrium models only after collecting data rather than applying pre-existing equilibrium models to available data. This third type of model is designed to explain that data but only after it is established that the statistical properties of the data are appropriate for the econometric model being estimated.[32] Presumably, some of those who engage in such model building activity think they are proceeding 'inductively' but that would be a misuse of the term 'inductive' in this case. The prior examination of the data is only about the econometric techniques used to estimate the parameters of the model and not about determining what to assume in the model.[33]

31. Which was discussed in Chapter 9 (see further Boland [2003, chs. 1, 5]). For a discussion of instrumentalism in the context of model building in general, see Boland [2014, ch. 11].

32. Readers who are interested in more about this third alternative are invited to read Boland [2014, ch. 10].

33. I discussed the misuse of 'induction' in Chapter 6 and will add more about this questionable matter of induction-based models in Chapter 12.

## CHAPTER 11

ᴄⱽɔ

# Equilibrium models vs. evolutionary economic models

It has long been recognized that one of the main alternatives to building general equilibrium models is provided by so-called evolutionary economics. Let us begin by considering the views of practicing evolutionary economists such as Richard Nelson, Sidney Winter and Ulrich Witt as to what they think constitutes evolutionary economics – basically evolutionary economics seems to reject equilibrium analysis. Nelson and Winter are usually seen as the earliest promoters of evolutionary economics and as they see it [2002, pp. 23–4]:

> [A]s one contrasts the economic textbooks and journals from prior to World War II with after, it is clear that while economics before the war still contained many evolutionary strands and concepts, these seemed to vanish in the early postwar period. What happened?
>
> The central factor, we believe, was the increasing fixation of neoclassical economic theory on equilibrium conditions ... and the mathematical formulation of that theory.... It became the standard view that microeconomic theory was about equilibrium conditions.

In Witt's 2008 *New Palgrave* entry on evolutionary economics, he observes,

> The question in evolutionary economics is ... not how, under varying conditions, economic resources are optimally allocated in equilibrium given the state of individual preferences, technology and institutional conditions. The questions are instead why and how knowledge, preferences, technology, and

institutions change in the historical process, and what impact these changes have on the state of the economy at any point in time.

A few years later, Nelson [2013, p.18] adds,

> Evolutionary economists ... might point out that, in an economy of continuing creative destruction that erodes much of the relevance of past experience, the standard assumption behind the demand curve concept ... and the assumptions behind the notion of a supply curve ... become very problematic. So does the assumption that the prevailing set of prices and quantities bought and sold are those of an equilibrium. More generally, the orientation of today's standard price theory is ill adapted for analysis of the determinants of price and price change in contexts where relatively radical innovation is going on and individual and collective learning processes are centrally involved in the action.

At the 1994 meeting of the Western Economic Association there was a session celebrating the 80th birthday of Armen Alchian. At this meeting several of his students and friends presented papers about Alchian's famous 1950 article, 'Uncertainty, evolution and economic theory'. Some of them claimed that his famous article was about evolutionary economics. Interestingly, Alchian was invited to respond to those papers. And as I remember, the first thing he said was that his article was not about evolutionary economics. About this I think he was right since all of his conclusions were about the consequences of reaching an equilibrium. The above quotations from practicing evolutionary economists illustrate what they think really constitutes evolutionary economics – basically it rejects equilibrium analysis. What Alchian's 1950 article was about was his response to the 1940s critics of the realism of the presumption of profit maximization employed by neoclassical economic model builders. A typical criticism was that people do not *consciously* maximize profit and hence that that presumption needs to be rejected in economic models.[1]

As to Alchian's 1950 argument, I would characterize it as arguing that conscious profit maximization by perfect competitors was not a necessary presumption for reaching a long-run equilibrium. Regardless of how the long-run equilibrium is reached, at that equilibrium all firms will be maximizing profit. This is because, if there were a way to reduce cost and thereby gain a competitive advantage, any perfectly competitive firm would have already taken it – and being a long-run equilibrium, there is no reason for anyone to change. Moreover, being perfect competitors they are all at their lowest level of average cost, and at that point, whenever long-run competition renders a zero profit it also means (by definition of average cost) that that price will

---

1. For examples, see Hall and Hitch [1939] and Lester [1946].

equal average costs. Average cost being lowest also means marginal cost will equal average cost and thus the producers are all maximizing profit since this also means marginal cost will equal price.[2] And most important, this will be the case whether or not they are consciously trying to maximize, as it is just a logical consequence of being at a state of long-run equilibrium.

Contrary to Alchian's claim that his article was not about evolutionary economics, in his note 7 he does say that he is combining Marshallian analysis with 'the essentials of Darwinian evolutionary natural selection'. I think his reference to Marshallian analysis is a reference to Marshall's equilibrium analysis. But, of course, contrary to Alchian's use of the evolution concept, evolutionary economists are among the critics of equilibrium models – and these critics are more associated with the older culture's perspective discussed in Chapter 5. For most of these critics, the concept of an equilibrium state is rejected outright. For others, a state of equilibrium is viewed at best as a temporary, static state that is not worthy of much interest.

It is important to keep in mind that evolutionary economic models are about the same economies that neoclassical economic models try to explain. Moreover, as Jack Vromen explains [2012, p. 739]:

(Static) equilibrium analysis is discarded in evolutionary economics and so are strong rationality assumptions. Agents are boundedly rational at most. They satisfice rather than maximize. What is more, agents, firms in particular, differ with respect to their behavioral properties. There is heterogeneity in this respect. Thus representative agent type of theorizing is rejected. So is equilibrium theorizing. There is no presumption that economies (or industries) are in equilibrium. There is no presumption even that economies tend to move in the direction of equilibria.

As I briefly mentioned in Chapter 5, evolutionary economists can be seen to be recent members of the 'older' culture who see an equilibrium state as something we should be able to see out our windows. In effect, they are never going to be satisfied with a state of equilibrium as being about something found only in a formal equilibrium model. In this chapter I will be discussing two basic schools of thought concerning evolutionary economics: an older school and a newer school. Despite the differences, we will see that both schools think economic models must recognize time and dynamics. Whereas the older school includes those that recognize time in a Darwinian manner, the newer school, the one that today dominates discussions of evolutionary economics, includes those that are more concerned with what is missing in

---

2. All of this was explained in note 12 in Chapter 8.

mainstream neoclassical equilibrium models including those elements identified by the critics I discussed in Chapter 5.[3]

## 11.1. DARWIN AND EVOLUTIONARY ECONOMICS

Darwin's view of evolution was a hot topic in the 1920s and 1930s.
Characteristic of this was the common notion of the 'survival of the fittest'.
Many saw this as a good representation of Darwin's view of evolution. This, of
course, was not at all what Darwin said about evolution. As Geoffrey Hodgson
notes in his 1997 article about Schumpeter's economics, this common characterization of evolution was due to Herbert Spencer's 1864 book about biology.
But, Hodgson also interestingly observes [1997, pp. 135–6]:

> Words like 'development' and even 'evolution' do not capture the essence of
> Darwin's theory. It was Herbert Spencer, not Darwin, who popularised the term
> 'evolution'. Darwin did not introduce the word until the sixth edition of the
> *Origin of Species*, and then only sparingly.

While Darwin's view was a hot topic, particularly in discussions of evolution
involving biological analogies, this was not true among all social scientists.[4]
Moreover, unlike Alchian, many economists (particularly Schumpeter) who
talked about the 'evolutionary process' or development of economies, did not
use the biological analogies often involved in notions of evolution.[5] Reflecting
on his contribution to evolutionary economics with Winter, Nelson [2001,
§1] adds:

> Let me begin by saying that I often have wondered if the body of theory that
> Sid Winter and I have tried to develop might have gone over better had we not
> used the term evolutionary. The term evolutionary carried a heavy biological,
> Darwinian, load. And yet, as I reflect on how I got into this arena, what was in-
> fluencing me was not Darwin, but Adam Smith, a great classical economist who
> came along before Darwin.

At the end of the nineteenth century, Thorstein Veblen, of course, did advocate making economics an 'evolutionary science'. But this did not seemingly attract attention among mainstream economists until the 1970s. There
Winter and Nelson published a well known article promoting their version

---

3. In the next chapter I will look at the Santa Fe Institute's research which could
easily be seen as a third school of evolutionary economics.
    4. See Degler [1991].
    5. See further Hodgson [1997, p. 134].

of evolutionary economics. Interestingly, in the published version of his PhD thesis, Winter [1964] critically examined Alchian's 1950 article with regard to connecting profit maximization with firms entering or exiting a market. But there is no mention of evolution or biology in his thesis. Such a mention was postponed to his 1971 article that was also critical of Alchian's 1950 article.[6] While Winter's later 1971 article does mention both biology and evolution, most economists would identify the 1974 article by Nelson and Winter as the beginning of what is called evolutionary economics today.[7] This is so even though they also constructed an equilibrium model. But, it is not a neoclassical model. As they say [pp. 887–88]:

> The time paths of output, input, and prices are interpreted as the paths generated by maximising firms in a moving equilibrium driven by changes in product demand, factor supply, and technological conditions. . . . The role of competition seems better characterised in the Schumpeterian terms of competitive advantages gained through innovation, or early adoption of a new product or process, than in the equilibrium language of neoclassical theory.

As Nelson suggested, in a 2001 talk (quoted above), there was no mention of Darwin in Nelson and Winter [1974] and no mention of biology, either. Their point was that once technology is considered an endogenous variable in one's model, that model can be considered to be an evolutionary model based on their claimed evolutionary theory of the firm. In particular they explain [1974, p. 891]:

> The first major commitment of the evolutionary theory is to a 'behavioural' approach to individual firms. The basic behavioural premise is that a firm at any time operates largely according to a set of decision rules that link a domain of environmental stimuli to a range of responses on the part of firms. While neoclassical theory would attempt to deduce these decision rules from maximisation on the part of the firm, the behavioural theory simply takes them as given and observable. The plausibility of this approach has, we think, been adequately established by previous work on the behavioural theory of the firm.

Despite this recognized beginning of modern evolutionary economics, most model builders in economics today think that an evolutionary model would be constructed analogously with evolutionary biology. But this is a

---

6. In Winter's 1971 article he does build an equilibrium model and in an appendix he provides a proof of the existence of competitive equilibrium under the assumptions of that model.

7. Other economists might instead credit a better known 1982 book by Nelson and Winter.

bit limited. If they were constructing their models to be analogous to evolutionary biology their models would be able to identify the four major building blocks of an evolutionary biology. In a 1994 article, Giovanni Dosi and Nelson specified these four building blocks as '(i) a fundamental unit of selection (the genes); (ii) a mechanism linking the genotypic level with the entities (the phenotypes) which actually undergo environmental selection; (iii) some processes of interaction, yielding the selection dynamics; and, finally, (iv) some mechanisms generating variations in the population of genotypes and, through that, among phenotypes' [p. 155]. Of course, as Alchian recognized, the competitive market seems most obviously an analogue of evolutionary biology's notion of environmental or 'natural' selection. However, evolutionary economics model builders do not usually speak in terms of the biological entities of 'phenotypes', but just think more in the limited terms of genes. In this regard, the most common notion of evolutionary dynamics is based on genetic mutations for which innovation seems to fill the analogical bill. The process of interaction is usually seen in terms of gene replication and in evolutionary economics gene replication is less obvious as one would find it difficult to think of an analogous form of biological reproduction between firms. How one would characterize replication usually depends on what is considered the economic analogue for genes.

While Nelson said he does not see their version of evolutionary economics to involve Darwin or biology, in their 1974 article Nelson and Winter did propose an analogical evolutionary role for a firm's internal routines or 'decision rules' that are usually the basis for management or production decisions.[8] And later in their 1982 book on evolutionary economics, they do go on to explicitly say 'In our evolutionary theory, these routines play the role that genes play in biological evolutionary theory' [Nelson and Winter 1982, p. 14]. In this regard, an innovation being represented as a change in such rules or routines – presumably the result of 'R&D'[9] – serves as the analogy for a genetic mutation. But, can such an analogue for genes and genetic mutations explain economic evolution?[10]

Vromen in relatively recent 2006 and 2012 articles has identified two supposedly different views of Darwinian evolutionary economics.[11] The two views are distinguished by what evolutionary models would presume about the role of biological evolutionary perspectives and analogies. They differ mostly

---

8. See Nelson and Winter [1974, p. 892].

9. To be clear, R&D is just the usual shorthand for Research and Development.

10. Note that Jack Vromen has argued that 'Even if we were to have complete knowledge of genes and skills, we would still fall far short of being able to predict the behavior of individuals, of routines and of firms' [2006, p. 545]. In any case, any evolutionary model would seem to need to include assumptions that characterize the necessary evolutionary biological elements or analogues.

11. See also Foster [1997].

concerning the role of the notion of 'routines as genes' that was promoted by Nelson and Winter [1982]. Vromen says that opposed to the Nelson and Winter approach are 'those questioning the appropriateness of "the biological metaphor" for the study of economic evolution' [2006, p. 544]. In particular, 'it was objected that there is no credible analogue of genetic inheritance in economic systems and that it is misleading to compare innovations to mutations, given that innovations can hardly be treated as random or blind processes' [ibid.]. The point, I think, is still that both alternatives recognize the essential ingredients of the Darwinian approach such as mutations, replication, selection, and so on. They just differ on how these essentials are used or represented in one's model.

## 11.2. NON-DARWINIAN THEORETICAL EVOLUTIONARY ECONOMIC MODELS

The non-Darwinian theoretical evolutionary models usually focus on missing elements in neoclassical equilibrium models. In many cases, equilibrium per se is not being rejected but focus is put instead on what happens to an equilibrium model when a missing element is instead included in the equilibrium model. As Dosi and Nelson explain in the 1994 article [pp. 154–5]:

> [L]et us first mention that we use the term 'evolutionary' to define a class of theories, or models, or arguments, that have the following characteristics. First, their purpose is to explain the movement of something over time, or to explain why that something is what it is at a moment in time in terms of how it got there; that is, the analysis is expressly dynamic. Second, the explanation involves both random elements which generate or renew some variation in the variables in question, and mechanisms that systematically winnow on extant variation. Evolutionary models in the social domain involve some processes of imperfect (mistake-ridden) *learning and discovery*, on the one hand, and some *selection mechanism*, on the other. With respect to the latter an evolutionary theory includes a specification of the determinants of some equivalent of a notion of fitness.

As Dosi and Nelson go on observe [1994, pp. 159, 166], if one were to repair a neoclassical equilibrium model by, in effect, following complaints like Hayek's and addressing the need to recognize learning, doing so can make the ultimate state of equilibrium 'path dependent'.[12] And this is even more important

---

12. Brian Arthur [1994] also suggested this. I will be discussing his views in the next chapter.

if satisfying those critics who think increasing returns must be recognized, since the usual reason for the existence of a firm's increasing returns is that the firm is learning how to increase productivity.[13] Unlike the presumptions of mainstream neoclassical competitive general equilibrium models, in which it is presumed that whatever was needed to learn was learned,[14] in evolutionary economics models perfect knowledge is not presumed and in many cases simply denied.

The difficulty with including learning in a typical neoclassical general equilibrium model is, again, that there is no optimum way to learn. And given this difficulty, neoclassical model builders are precluded from relying on the neoclassical behavioural assumption of optimization when explaining the outcome of learning. Moreover, as Dosi and Nelson note in the quotation above, during the process of reaching the long-run equilibrium, the firm is influenced by random events which make the eventual long-run equilibrium path dependent (on those events). No two firms even when starting from the same state of technology would likely reach the same long-run equilibrium state.[15]

The classic example of path dependency is the one discussed by Paul David in a 1985 article about the non-optimal QWERTY keyboard we all use today. There are, supposedly, better keyboard arrangements of the keys. One is the Dvorak keyboard with which, if you knew how to use it, you could increase your typing speed significantly. The reason why we are stuck with the supposedly inferior keyboard has to do with the history of typewriters. The original keyboards were mechanical and when the typist tried to type too fast, the keys would jam and the typist would have to stop to separate the keys.[16] The

---

13. This has a lot due to the Marshallian view of the firm's production function. It is as follows: (1) it begins in Stage I with a rising positive marginal product (of inputs such as labour) as this firm grows in size (by increasing inputs); (2) in Stage II the marginal product stops rising and begins falling; (3) and eventually in Stage III marginal product falls to zero which means the total output reaches its maximum after which if the input were to increase, output would begin falling.

These three stages of a production function correspond to the old Victorian notion of a 'three-generations' model of a firm: Grandpa who created the firm is learning in Stage I (hence increasing returns and the associated rising marginal product). When the firm is passed on to the son, the son operates in Stage II, in which output is rising but at a diminishing rate (this stage is the main subject that Economics 101 textbooks talk about in their chapter about the supply curve of the firm). When in Stage III the son passes the firm onto the grandson, the firm begins to fail (which is not often discussed in Economics 101) and in this stage, the marginal product is thus negative.

14. And thus the presumption of perfect knowledge that critics complain about.

15. Unless they deliberately try to match each other firm, of course.

16. For young readers who are familiar only with computer keyboards, let me explain. On the old mechanical typewriters, each letter was molded in reverse on the end of a long thin bar which produced the image by striking the molded reverse key on the inked ribbon over the paper. The problem arose because the individual bars with the reversed letters were mechanically connected to the corresponding keys on the keyboard. Pushing on a key would cause the letter to swing a significant distance to strike

solution to this problem was to recognize which keys were used most often and then use that information to rearrange them on the keyboard to make it difficult to cause the conflict when keys are quickly pushed in succession. The QWERTY configuration was the resulting solution. When the mechanics of the typewriter were eventually made to prevent a mechanical conflict, there was no need for the QWERTY keyboard but it remained. The primary reason for the QWERTY retention is that once we all learned to use the QWERTY keyboard without having to think about where the keys are, changing to something like the Dvorak keyboard would require everyone to learn to type all over again. In other words, today's equilibrium keyboard (QWERTY) was reached by way of a particular historical path and not because it was chosen as a universally optimum keyboard configuration. And QWERTY remains, of course, even though we eventually began using electric typewriters such as the IBM Selectric typewriter which can never involve conflicting keys.[17]

Dosi and Nelson also identify other cases of path dependency – most due to the introduction of new technologies. In particular, regardless of where a firm's current technology came from, the dynamics of increasing returns due to learning to use that technology preclude easy changes to a new technology which after a sufficient amount of time might prove to be a better technology. Learning and the dynamics of increasing returns leads to a certain inertia that precludes easy adoption of optimum technologies.

The main point of the discussion of technologies here is the recognition that the technologies being used are endogenous variables and not the exogenous variables often presumed in the typical neoclassical or Walrasian general equilibrium model. And this endogeneity makes the idea of a general evolutionary model difficult to envision. Instead, evolutionary models tend to be case by case studies as with the QWERTY keyboard although, as in the case of learning or increasing returns, there are some basic strategies to consider.

## 11.3. ALTERNATIVE VIEWS OF EVOLUTIONARY ECONOMIC MODELS

While some evolutionary economists are interested in addressing various theoretical problems without any concern for building formal models, it is

---

the ribbon; pushing on two adjacent keys too quickly in succession allowed their letters to strike each other, resulting in a jam.

17. The speed of the Selectric typewriter by design was not dependent on the keyboard's configuration and thus could never have the problem that the earliest mechanical typewriters had, since all of the letters that were once located on separate bars were now all on one ball (a typeball) that rotated in response to electronic signals sent by a keys on the keyboard.

important to recognize that Nelson and Winter [1974] did include a formal model and today some of those involved with evolutionary economics are also interested in building formal models. It should be noted that most who are interested in building formal evolutionary economic models are concerned with the dynamics within individual firms and how the dynamics change through time. It should be noted that such models are also relevant to behavioural economics and even to sociology. As discussed in Section 11.2, evolutionary models often address such basic things as a firm's technology but, going further, this could also be extended beyond the firm to address such basic things as consumer preferences and beliefs, that are presumed to have been exogenously given. Some model builders have been going so far as building formal models to show that consumer preferences might have been affected by biological evolution.[18]

It should, of course, be recognized that there are those evolutionary economists who are less concerned with building formal models – particularly those who think we should try to understand economics before we resort to equilibrium-based explanations.[19] Moreover, as discussed in Section 11.1, it should be recognized that one still does not have to adopt a Darwinian perspective to build an evolutionary model. A relatively recent example that specifically addresses evolutionary modeling without relying on a Darwinian perspective can be found in a 2012 article by Kenneth Carlaw and Richard Lipsey. Carlaw and Lipsey [p. 736] argue that economists face two 'distinct paradigms' – yielding two 'conflicting visions of the market economy'. One vision sees the behaviour of the economy being the result of reaching an equilibrium between opposing forces. There, the market's dynamics are characterized by negative market feedback and ends with the economy returning to its static equilibrium.[20] In the other vision, the behaviour of the economy is the result of many different forces including various technological changes. Such changes evolve endogenously over time as they involve positive market feedback. Carlaw and Lipsey distinguish the two visions by the *stationarity* in the former view and *non-stationarity* in the latter view. What is important here is that Carlaw and Lipsey are clearly stressing that the Darwinian-type evolution is not the only way to characterize non-stationarity.[21]

---

18. For example, see Robson [2002].
19. See Hodgson and Knudsen [2006], Witt [2008] and Vromen [2012].
20. Or just returning to its stationary equilibrium growth path.
21. Note also that they actually have provided an explicitly non-Darwinian evolutionary model in Carlaw and Lipsey [2011].

## 11.4. GOING BEYOND THE EVOLUTIONARY THEORY OF THE INDIVIDUAL

While, as I noted earlier, the focus of much of Nelson and Winter's evolutionary work is on a behavioural microeconomics model of an individual firm, one can adopt a more macroeconomic view of evolutionary economics. One might address the sociological aspects of where consumer preferences come from. This can be similar to what one can find in much of the computer-based evolutionary game-theoretic models – such as those about population dynamics. It might even be claimed that since evolutionary economics is usually about change, it should be seen as a challenge to the old Walrasian general equilibrium perspective on the macro economy. One standard complaint about Walrasian general equilibrium models is their lack of consideration of heterogeneity of agents in a macroeconomic economy such as the diversity of tastes and abilities.[22] In this regard, the macroeconomic perspective would convince some evolutionary economists that evolutionary models do not actually need to delve into biological metaphors or analogues.

In this regard, almost every evolutionary model builder thinks that a main issue that needs to be explained is the basic notion of selection, since that is the basis for most explanations of evolutionary change. But, must change always be considered the result of a process analogous to genetic-replication-based selection or is it the result of environmental interaction or perhaps both? Some formal evolutionary economic model builders have turned to the work of the geneticist George Price [1972, 1995]. In particular, his work of interest to evolutionary economists is an equation called the 'Price Equation'.[23] As Thorbjørn Knudsen [2004, p. 155] has noted:

> The Price Equation is useful for conceptual as well as empirical purposes, and it is consistent with previous approaches in evolutionary economics. For conceptual purposes, the Price Equation can be used to distinguish between the bare minimal requirements of a valid evolutionary explanation.

The Price Equation can also be shown to be consistent with the work of Nelson and Winter, particularly, their formalization of evolutionary economics. Nevertheless, I think it is open to question whether the Price Equation is useful for modeling economic evolutionary dynamics. Moreover, if we try to test a model of evolutionary economics that is modified by adding a formal equation such as that which Price suggested and the model is empirically

---

22. See Boland [2014, ch. 2].
23. See Knudsen [2004, p. 154].

tested and refuted, we do not know whether the refutation is due to this additional equation or the original unmodified model's assumptions about evolutionary change, which would be our primary reason to test the model.[24] Perhaps a more important question that should be addressed to these formal evolutionary model builders is: why do they think there is a need to build formal models?

24. This is a well known logical problem of testing models – the Duhem-Quine problem – and is extensively discussed in Boland [2014, ch. 8]. But what is important here is that this problem has nothing to do with formal evolutionary models not being equilibrium models. It is a problem facing the testing of all formal models consisting of multiple assumptions.

# CHAPTER 12

# Equilibrium models vs. complexity economics

In the mid-1980s, a small group of researchers including economists and physicists gathered together at an Institute in Santa Fe, New Mexico to try looking at the state of economic understanding, among other things, and to see what could be done to advance our understanding of economics and maybe even of physics. At various times, the economist Brian Arthur headed the group and he began publishing articles about what he called 'complexity economics'.[1] What these researchers began talking about was to a great extent what the non-Darwinian evolutionary economists were talking about for many years. Mostly it was about dealing with all of the problems with Walrasian general equilibrium models identified by the critics that I discussed in Chapter 5.

## 12.1. COMPLEXITY ECONOMICS

Let us begin by looking at the published views of Arthur concerning the nature of complexity economics. In the preface to his book that includes a collection of his work on complexity economics, he says [2014, pp. xix–xx]:

> [T]he features of complexity economics are clear. The economy is not necessarily in equilibrium; in fact it is usually in nonequilibrium. Agents are not all knowing and perfectly rational; they must make sense of the situations they are in

1. Most of these articles have been recently published as a collection in Arthur [2014].

and explore strategies as they do this. The economy is not given, not a simple container of its technologies; it forms from them and changes in structure as this happens. In this way the economy is organic, one layer forms on top of the previous ones; it is ever changing, it shows perpetual novelty; and structures within it appear, persist for a while, and melt back into it again. All this is not just a more poetic, humanistic view of the economy. It can be rigorously defined, and precisely probed and analyzed.

From page 2 of the first chapter, he says:

> [T]his new approach is not just an extension of standard economics, nor does it consist of adding agent-based behavior to standard models. It is a different way of seeing the economy. It gives a different view, one where actions and strategies constantly evolve, where time becomes important, where structures constantly form and re-form, where phenomena appear that are not visible to standard equilibrium analysis, and where a meso-layer between the micro and the macro becomes important. This view, in other words, gives us a world closer to that of political economy than to neoclassical theory, a world that is organic, evolution-ary, and historically-contingent.

And on the next page he adds:

> Complexity is not a theory but a movement in the sciences that studies how the interacting elements in a system create overall patterns, and how these overall patterns in turn cause the interacting elements to change or adapt. . . . To look at the economy, or areas within the economy, from a complexity viewpoint then would mean asking how it evolves, and this means examining in detail how in-dividual agents' behaviors together form some outcome and how this might in turn alter their behavior as a result.

With this in mind, it is important to recognize that the beginning researchers at the Santa Fe Institute all understood modern mainstream economics as the study of an economy by building formal models with simultaneous equations for which we would explain prices by solving such models. The equations, as in Walrasian general equilibrium models, represent individual agents (demand-ers and suppliers) that, in the words of Voltaire, 'till their own gardens'. In the mainstream perspective one does not consider interactions between market participants. Judging by Economics 101 textbooks, usually the model build-ers do not seem to consider how institutions that support the market come about. And certainly, they do not consider how institutions might be changed by the participants in the market beyond bidding prices up and down with the resulting transactions. As discussed in Part I, to claim the existence of an equilibrium one must also explain how that equilibrium is reached, if a state of equilibrium is to be the foundation for an explanation of any aspect of the economy.

What the group at the Institute recognized is that if one tries to build a model to deal with the various aspects of agents socially interacting in an economy, the mathematics of that model will get very complex – hence the notion of *complexity* economics which Arthur is explaining in the above quotations. Part of the reason for why social institutions might change over time is, of course, that everyone is learning from what takes place as a result of the essential sociological interactions.[2] As noted in Chapter 11 (Section 11.2), once we start including learning in an equilibrium model we need to recognize that while one can conceive of an optimal demand or supply choice, there is no optimal way to learn. And the lack of an identifiable optimal way to learn means that anything can happen – which makes an eventual equilibrium state path-dependent. That is, random mistakes and random events in the learning process mean that one cannot predict the eventual state of equilibrium as one can always do with Walrasian general equilibrium models; solutions to these models amount to predictions that would be confirmed if the models were actually about the real world. This is because a solution to a Walrasian general equilibrium model is the result of everyone obtaining their *unique* optimal (i.e., maximizing) choices. Of course, there still is the matter of technological innovations which, by definition, go beyond what is available and understood by participants in the economy. For the old Darwinian inspired evolutionary economists, the idea of an innovation was employed to analogically represent a gene mutation found in biological evolution. For the complexity economists, what happens after an innovation is what will involve evolution.

Unlike the rational expectations theorists of the 1970s and the Real Business Cycle theorists of the 1980s, the complexity theorists do not attribute the dynamics of an economy to exogenous changes. Instead, they focus on the endogenous reasons for change and thus they focus on 'non-equilibrium' models. Any apparent equilibrium that might be possible is only one of many possible situations.

The question of course is, what do we do if we are not building equilibrium models? Arthur's [2014, pp. 8–9] suggestion is to see the workings of a non-equilibrium economy as something like a computer program. He sees this change of perspective as the economy becoming 'algorithmic'. Presumably, instead of building an equilibrium model as a collection of simultaneous equations for which we try to solve for the variables we are using the equilibrium

---

2. See also, Boland [1971, 1979b]. These two articles were written after teaching a class in introductory sociology about social institutions. Based on what I learned in that class, I wrote a paper [1979b] that argued that institutions exist in two forms: consensus and concrete. In both cases, they exist to preserve social knowledge about how to solve recognized social problems.

model to explain, we would write something like a computer program by specifying a set of algorithms – a set of 'if A then do B' statements – and 'run' the program to see what happens. If successful, it presumably reaches a dynamic setting that matches what is going on in the economy being explained. Of course, it could end in a state of equilibrium but that is only one of a many possible ongoing states.[3] In short, instead of building equilibrium models, complexity economics involves building non-equilibrium models in the form of computer programs.

## 12.2.  TECHNOLOGY, INCREASING RETURNS AND EVOLUTION

One can easily see how the running of the computer-program model of the economy can be seen as the economy being modeled to demonstrate the evolution of that economy. There are many important elements of an economy that can be incorporated in the computer-program model, such as increasing returns and endogenous changes in technology. As Arthur notes 'Under increasing returns, competition between economic objects – in this case technologies – takes on an evolutionary character ...' [1989, p. 128]. Increasing returns has an old history but, as I have already noted, the old version was related to the notion of new firms learning to do their business and in particular, learning to lower production costs. That is not actually the same thing that Arthur has in mind. His version is more directly related to changes in technology. Of course, learning to use a new technology is a bit like the old version of increasing returns but that would more be the case when the change in technology is exogenous. Exogenous change is not what Arthur and members of his Santa Fe group usually have in mind.

The basic notion of an endogenous technological change for the Santa Fe group is that a new technology is often a combination of old existing technologies.[4] And it is the new combination that gives rise to the increasing returns that in turn creates an opportunity for learning how to more efficiently use the new combination. This notion of a technology change would seem to fit well with a computer-program model. The program can tryout various combinations; throw in some random variations or perhaps have the program look for problems that such a combination might solve. All of this would be something like an evolving economy.

---

3. Interestingly, unlike the equilibrium-model methodology which usually needs exogenous shocks to produce business cycles, the computer-model approach would allow for bubbles and crashes as in the case of their 'Santa Fe stock market model'.
4. See Arthur [2014, pp. 19–21].

## 12.3. DIVERSITY, LEARNING, PATH DEPENDENCY AND EVOLUTION

One of the complaints about the DSGE models that include a representative agent is that the real world exhibits considerable diversity which is lost in such equilibrium models. If the representative agent is relied on, it means all agents represented are in effect responding to market signals in exactly the same way. Diversity at best is simply ignored. The Santa Fe computer-program model does not have to ignore diversity. Diversity with a little randomness can easily mean the running program is exhibiting something like an evolving economy.

In such an evolving economy, randomness plays a role in learning. Specifically, if there is learning in a model (including any general equilibrium model, by the way), then random events in the process of learning can mean that the outcome may be path-dependent as Dosi and Nelson note '[If] one bases rational choice theory on accumulated learning, there are apparent limitations to the explanatory power of the [neoclassical] theory even in these cases. In particular, learning processes may be very path dependent. Where they end up may depend to a considerable degree on how they got there' [1994, p. 158]. The learner might go one way because of the random event but some different random event might cause the learner to go a different way. This is obviously contrary to solving a general equilibrium model for a unique set of equilibrium prices.

Surely, though, not all new technology is a combination of preexisting technologies. Where do really new technologies come from? I suspect that one can easily criticize the Santa Fe computer-program modeling by noting that if truly new elements of the economy are recognized, the programming-based evolutionary model of the economy is limited to only discussing some historical event or events.[5]

## 12.4. LEARNING AND 'INDUCTIVE REASONING' IN SANTA FE: A CRITIQUE

I suspect that the response of the Santa Fe group to my critical observation that there are limits to computer-program-based modeling would be to suggest that their program could engage in producing new ideas by engaging in

---

5. I remember Herbert Simon giving a talk in the 1980s at the Canadian Economic Associations annual meetings. The talk was about an artificial intelligence model of a famous old physics experiment. His computer program duplicated what happened or at least the eventual outcome of the experiment. I also remember thinking that such an exercise was not demonstrating the discovery of something new – it was just demonstrating a history. Perhaps the Santa Fe models will always have the same problem.

what they call 'inductive reasoning'. As I will explain in this section, Arthur and perhaps other members of the Santa Fe group are profoundly confused about induction,[6] inductive reasoning and rationality (bounded or otherwise).

As I noted in Chapter 9, economists throw the word 'rational' around without ever giving much thought to what it supposedly means, or much thought to its use in the eighteenth century.[7] And as I explained, rationality is only a property of arguments – it is not some psychological process. A rational argument is a logically valid argument. A logically valid argument is one for which, whenever all of its assumptions are true, all logically deduced statements from that argument must be true. If economists were more careful they would recognize that when they say a consumer is rational, they merely mean that it is possible for the economic theorist to *explain* the choice that the consumer makes. The economic theorist does this by providing a logically valid argument for which one can validly deduce a statement that represents the choice made.[8] Such an explanatory argument consists of assumptions, at least one of which is of a universal nature (e.g., *all* people are maximizers) but the argument must also have statements of a non-universal nature (e.g., statements that might identify that consumer's preferences and non-universal observation statements about the consumer's budget as well as the prices faced by that consumer). Because we can do this, we are supposed to say that all the consumers are rational but this claim is simply false: it is only the provided argument that is rational.[9]

---

6. This is another term economists seem confused about and involves the one I briefly explained in note 22 of Chapter 1. There I discussed the notion of induction that the eighteenth-century philosophers were relying on. They thought there is an inductive logic for which it was presumed that that form of logic not only existed but with it one could reliably proceed from observations alone and inductively prove some general proposition. Unfortunately, as I have explained, there is no such logic and we have known this for some time. After all, David Hume proved its impossibility in the late eighteenth century. The question then was whether knowledge can be based only on experience, to which a philosopher like Bertrand Russell would ask, 'how do you know knowledge can be based only on experience?', to which, if you were consistent, you would have to answer 'by experience alone' which leads to an infinite regress. So, as a matter of logic alone, inductive learning is impossible. Unfortunately, some economists think any argument which includes premises in the form of observational statements is thereby an inductive argument. But this presumes that a deductive argument cannot involve such statements. And this is the key mistake. After all, initial conditions are of the form of observational statements. Again, see Boland [2003, ch. 1] or, for a general discussion of induction in model building, see Boland [2014, ch. 11].

7. See again note 6 of Chapter 9.

8. But be careful about this. A logically valid argument may be false and will be false if any of its assumptions are false. All that logical validity assures is that the deduction must be true *if* all the assumptions are true. Boland [2014, p. 244] provides a logically valid argument for which the deduction is true but the assumptions are false and hence the argument is thereby false.

9. Of course, some unusual consumers might actually engage in such a tedious exercise themselves but it is doubtful when talking about ordinary daily purchases such

Unfortunately, Arthur and some of the other members of the Santa Fe group seem confused about this. Arthur claims 'rigorously speaking there can not be deductively rational behavior' and he goes on to say [2014, p. 6]:

> One way to model this is to suppose economic agents form individual beliefs (possibly several) or hypotheses ... about the situation they are in and continually update these, which means they constantly adapt or discard and replace the actions or strategies based on these as they explore. They proceed in other words by induction.

Of course, the notion of induction has a long history. As Keynes long ago explained, 'Induction is in origin a translation of the Aristotelian ἐπαωγή. . . . The modern use of the term is derived from Bacon's. Mill defines it as "the operation of discovering and proving general propositions"' [1921, p. 315]. What does 'by induction' mean to the Santa Fe group? For that matter, what does 'inductive reasoning' mean? I suspect that the theorists at the Santa Fe Institute presume all rational arguments consist of assumptions that are universal or non-empirical statements (like 'all consumers are utility maximizers') as if the argument contains nothing representing an observation or any non-universal statement. But this presumption can easily be false. One might actually observe the prices faced by the consumer, or maybe even the consumer's budget, and as such the argument would be including an empirical non-universal statement. And so, I suspect that they raise this issue merely because they want to make sure observations matter. The bottom line of my criticism here is that all explanatory arguments include both universal and non-universal statements.[10]

But the most important point to keep in mind is that contrary to what they seem to be suggesting: there is no inductive rationality; there is only deductive rationality. To be a *purely* inductive argument would require that the only statements allowed to be included in the argument would be singular observation statements, and furthermore, to complete the argument would require an impossible infinity of observations. This is because, for an inductive argument to be logically valid, the argument is required to prove the *impossibility* of any conceivable observation that would refute the argument.[11] Purely inductive arguments are simply impossible.

---

as a quantity of tomatoes that a point on the tomato demand curve represents. If they do it at all, it would usually be after the choice was made – and likely only if they are asked why they made that choice.

10. A trivial example is the syllogistic explanation for why Socrates died: It was because (1) *all* men are mortal (a universal statement) and (2) Socrates was a man (a non-universal statement) therefore (3) Socrates died.

11. The typical example of this would be about inductively proving the universal statement 'All swans are white' is true using *only* observations of white swans. Since the statement is unlimited in time and space, there is potentially an infinity

It turns out that the notion of induction being an alternative to deduction has a long and confused history. Let me explain for those who have read economists making similar claims for an inductive reasoning as those made by the Santa Fe Institute researchers. If one reads about induction or inductive reasoning in celebrated philosophy literature – such as Bertrand Russell's [1945] famous *A History of Western Philosophy* – one usually gets a very different impression than one will get by reading what some economists say about induction. Part of the difference might be explained by noting that even the economist-philosopher David Hume struggled to deal with induction.[12] He is credited with demonstrating that the common notion of 'induction by simple enumeration' was a logical impossibility. Even Francis Bacon, the seventeenth-century advocate of inductive method, understood the limits of 'induction by simple enumeration'. Bacon thought there was a better way to do induction – by simply arranging carefully obtained observational data such that the right explanatory hypothesis would be obvious.[13] But so far nobody has come up with a method that would reliably do this – some philosophers call this the 'problem of induction'.[14] Yet Hume maintained that people do inductive reasoning all the time – in their heads, so to speak.

So, again, what do the economists such as those at the Santa Fe Institute mean by induction or inductive reasoning? Surely, they do not mean 'induction by simple enumeration'. When economic theorists say people learn inductively do they mean that people make careful observations and inductively infer what are their preference maps or their production functions?[15] Are Santa Fe theorists thereby claiming people have a solution to the philosopher's problem of induction, or are they merely agreeing with Hume that, while there

of observations of white swans if the statement is true. But the question always remains: will the next swan observed be white, also? The inductive proof would have to prove that that next swan *will be* white and not some other colour. In short, an inductive proof would have to prove a non-white swan does not and cannot exist anywhere or at any time. Of course, in this particular case, we know this universal statement is false because one can go to the Westin Hotel in Kaanapali on Maui and observe a black swan floating in their pond!

12. As I discussed in Chapter 6.

13. Presumably by identifying or recognizing patterns in the data.

14. Robert Sugden [1998] explains three different problems of induction but all are problematic for economic modeling.

15. Stephen Spear specifically advocated induction as a means of explaining how people form rational expectations. In support of this he says [1989, p. 891]:

> Inductive learning is characterized in psychology as the process whereby new information from the environment is organized into cognitive patterns which then serve to either reinforce an individual's cognitive models of the environment, or to alter these models to conform with newly recognized patterns.

Clearly this is not what Bacon was talking about – but, as we shall see, Bacon was talking about something that today might still be called pattern recognition.

is no solution to the problem of induction, people can still inductively reason in their heads? The idea that they can do it in their heads seems to be the foundation of what theorists associated with the Santa Fe Institute mean by induction and inductive reasoning.[16]

The reason for economists discussing induction and inductive reasoning at the Santa Fe Institute seems to be that it could be a promising way to develop economics along lines which recognize the need to explain *how* individuals make decisions.[17] That is, for example, *how* do individuals make their maximizing choices? What do they need to know and how do they know it?

The thrust of the Santa Fe program seems tied to Hume's dodge of his problem of induction[18] – that is, regardless of the impossibility of fulfilling Bacon's program, we should see induction as a manageable psychological problem of the mind. Judging by the frequent references to the book titled *Induction* – a 1986 book by John Holland, Keith Holyoak, Richard Nisbett and Paul Thagard (all four are psychologists or researchers in connected fields) – it would seem that this book is the primary basis for the Santa Fe research concerning induction. What that book advocates is that we should try to understand how people learn by applying a form of Charles Peirce's so-called 'abductive' logic. If you watch old television crime or detective programs such as *CSI* or *Law and Order*, you will be familiar with this logic. Supposedly, if you show the suspect has the *means*, the *motive* and the *opportunity*, then one is well along the way to proving that the suspect is *guilty*. Logicians, however, will point out that this is an invalid mode of reasoning as it still has not proven that innocence is impossible. But Holland and his colleagues consider such lines of reasoning to be 'processes of induction' which they think people apply to practical

16. In an e-mail correspondence of 12 September 2015, Duncan Foley explained why my criticism here may be ill founded. Specifically, he explained what those of the Santa Fe group mean by 'deductive' and 'inductive': 'They want to contrast models that start from some theory of asset pricing from fundamentals and assume that agents are operating with some version of those theories, which Arthur would call a "deductive" or possibly "top-down" approach, with their models in which agents shift among algorithms of many types purely on the basis of their empirical performance, which they tend to call "inductive". I agree with your critique that this appropriation of language is unfortunate because of the massive confusion it is likely to create, but if you want to come to grips with the substance of the Santa Fe "vision" I think you need to look beyond the semantics to the substance of their ideas about how markets work'. I think he is in a better position than I am to understand what members of the Santa Fe group have been doing and so he is probably right about this. But given this, I would suggest to the Santa Fe researchers that they could easily avoid such confusion by not using such troublesome and probably unnecessary terminology such as 'inductive reasoning'.

17. Particularly in dynamic or disequilibrium situations – see Waldrop [1992, p. 325].

18. It may be unfair to Hume to call it a dodge since he was not claiming that one could justify one's knowledge psychologically, just that one could not justify one's knowledge claims by alleged inductive logic.

situations of problem solving.[19] The program, following Holland, et al., is then to introduce models of the mind that would explain how induction works.[20]

Like Bacon's view of science, the Institute's research in economics seems clearly to disfavour deductive explanations. To illustrate, consider the explicit view of Arthur who says [1994, p. 406]:

> The type of rationality we assume in economics – perfect, logical, deductive rationality – is extremely useful in generating solutions to theoretical problems. But it demands much of human behavior, much more in fact than it can usually deliver....
>
> Economists, of course, are well aware of this. The question is not whether perfect rationality works, but rather what to put in its place. How does one model bounded rationality in economics? Many ideas have been suggested in the small but growing literature on bounded rationality; but there is not yet much convergence among them. In the behavioral sciences this is not the case. Modern psychologists are in reasonable agreement that in situations that are complicated or ill-defined, humans use characteristic and predictable methods of reasoning. These methods are not deductive, but *inductive*.

He goes on to explain how we think inductively [pp, 406–7]:

> Modern psychology tells us that as humans we are only moderately good at deductive logic, and we make only moderate use of it. But we *are* superb at seeing or recognizing or matching patterns – behaviors that confer obvious evolutionary benefits ... [W]e look for patterns; and we simplify the problem by using these to construct temporary internal models or hypotheses.... We carry out localized deductions based on our current hypotheses and act on them. And, as feedback from the environment comes in, we may strengthen or weaken our beliefs in our current hypotheses, discarding some when they cease to perform, and replacing them as needed with new ones. In other words, when we cannot fully reason or lack full definition of the problem, we use simple models to fill the gaps in our understanding. Such behavior is inductive.[21]

19. Surprisingly, this is similar to what Popper [1966, ch. 14] advocates when he conjectures the problem situation and applies his 'situational logic' (which is a general form of neoclassical economic methodology – see Hands [1996] and Boland [2003, ch. 15]). But for some reason Holland et al. feel the need to excoriate Popper – could it be because Popper agreed with Hume's rejection of the empirical possibility of Baconian inductive method? Maybe it is instead that Popper rejects any Humeian retreat to psychology as a means to an inductive justification of one's learning or knowledge claims.

20. It should be recognized that by using a model of the decision maker's mind one is thereby *deducing* the decision maker's actions while within the model the decision maker is supposedly *inducing* the knowledge needed for us to explain those actions!

21. What Arthur describes here has nothing to do with induction. It is simply a case of trial and error such that each trial is a test of a prior conjectural assumption, not a prior observation.

To justify the need to study the uncertain or complicated situations faced by decision makers in economics Arthur, like many economic theorists, usually calls upon the notion of bounded rationality – but for him it is Thomas Sargent's, not Herbert Simon's.[22] But, like many Santa Fe projects, Arthur's study is nevertheless rooted in a psychology of the mind which, according to Holland et. al., is primarily about pattern recognition. Interestingly, this sounds like Bacon's process of induction but the Baconian approach had nothing to do with the mysterious workings of the human mind but simply about human experience.[23] It is such experience that Hume denied as an adequate logical basis for knowledge justification or for learning, and this denial led to his advocating the psychology-based approach instead. It would seem to me, then, that the researchers of the Santa Fe Institute may be advocating *both* Bacon's and Hume's view of inductive reasoning.

It is unfortunate that Arthur and his Santa Fe Institute colleagues find it necessary to show that they reject 'deductive rationality'.[24] Fortunately, such a rejection is not necessary for their promising computer-program model approach to explaining the evolution of a non-equilibrium economy. I think one can easily engage in such modeling without denying deductive rational argumentation which is, of course, the basis of all logically valid explanations including those found in economics – as well as in computer programs!

22. I explained this notion in Chapter 6, note 4.
23. Bacon was more concerned with people (quacks?) that would invent elaborate deductive arguments to prove their prejudices. For him, the solution to this was to encourage people to look at the facts of experience.
24. For example, see the above quotation from Arthur [2014, pp. 5–6].

# PART THREE

# Avenues for overcoming the limits of equilibrium models

## *Some methodological considerations*

# CHAPTER 13

⚭

# Building models of price dynamics

In this and the next two chapters I will be considering what I think should be learned from the discussions in Parts I and II. In doing so, I will occasionally be presenting a methodological examination of some of the major topics discussed there.[1] The main methodological topic that will be further examined is one that has been discussed already, namely, the requirements of methodological individualism that are taken for granted by mainstream equilibrium model builders and that raises the following questions that Fisher [2003, p. 77] asked:

> Whose behavior does [the price adjustment] equation[2]... represent? It cannot reflect directly the behavior of the individual agents whose demands are to be equilibrated. Indeed, we now see a central conundrum: In a perfectly competitive economy, all agents take prices as given and outside of their control. Then who changes prices? How do sellers know when demand or costs rise so that they can safely raise prices without losing all their customers? At a formal level such questions are deep ones.

For this purpose let me begin by returning to where I began in Part I, particularly the theoretical equilibrium modeling problem presented by Arrow

---

1. This examination will be concerned only with what is sometimes called 'small-m' methodology questions. Such questions are distinguished from the kind of high-level, sophisticated methodology issues that analytical philosophers would want to talk about. Here, as small-m methodology, I will be concerned only with low-level issues concerning how economists look at things and in particular I will be more concerned only with why economic model builders assume what they assume.

2. Specifically, equation 2.4 in Chapter 2 which is the same equation Arrow used in his 1959 article.

in his 1959 article. Here I will apply what I hope has been learned so far from my critical evaluation of research programs based on Arrow's theoretical challenge – some of which were discussed in Part II. Recall that Arrow said that our mainstream microeconomic theory of perfect competition explains an individual's behaviour by presuming the individual is a price taker and moreover by presuming that the prices the individual faces are equilibrium prices. Of course, this is the explanation that one will find in the usual mainstream microeconomics textbooks. And I think Arrow's article demonstrates that our textbooks' microeconomic theory is at best incomplete and at worst a contradiction. If one wishes to complete the theory of the behaviour of all individuals who are presumed to be equilibrium-price takers, Arrow said one needs to explain the process by which prices are adjusted to their equilibrium values.

As was discussed in Chapter 2, Arrow suggested that the most obvious explanation of price adjustment would be based on the textbooks' theory of an imperfectly competitive firm. A firm is considered imperfectly competitive whenever the number of firms in its industry is thought to be relatively small and thus any change in any firm's supply output is thought to affect the market price – which means then that each of these firms is actually facing a downward sloping demand curve. But since there is a small number of firms, each firm must take account of the effect its supply decision has on a market price that can vary with the supply decision rather than facing just one singular going price on which its supply decision has no effect. Explaining prices using an imperfectly competitive firm begs the question of how a firm knows the entire demand curve it faces. This, of course, was discussed in Chapter 3 where Clower's 1959 article was shown to demonstrate that, without the imperfect competitor having complete and perfect knowledge of the demand curve, assumptions must be made by the firm about the nature of that demand curve – but unless those assumptions are true, it is possible that the eventual equilibrium status of the market may be sub-optimal; and this is contrary to the textbooks' usual purpose for building equilibrium models based on the behavioural assumption of profit maximization. And, of course, this purpose has been pursued by equilibrium model builders since the eighteenth century, when Adam Smith in effect built his non-formal equilibrium model. A few economic theorists have long ago interpreted the process of equilibrium attainment correctly to be a matter of learning methodology[3] along the lines suggested by Hayek in his 1937 article. Unfortunately, most economic theorists have viewed Arrow's problem and his solution only as one of deciding what else to assume when building a formal model of the market equilibrium.

3. For example, see Gordon and Hynes [1970].

## 13.1. THE ANALYTICAL PROBLEM OF PRICE ADJUSTMENT AS PRESENTED BY ARROW IN 1959

Much of the discussion and criticism from both the older and the newer cultures identified in Chapter 5 were about the price dynamics that must take place *before* an equilibrium state is obtained.[4] Equilibrium model builders who want to explain only the equilibrium price will simply build Arrow's type of formal model by assuming equations much like those Arrow identified. Those equations are intended to represent how the total market quantity demanded and the total market quantity supplied depend on the price that is presented to the individual demanders and the suppliers. With such a model, the microeconomic textbooks would claim to have explained the price by assuming that the market is cleared. Some might engage in explicitly solving the model for the market-clearing price to show that market clearance is possible given the assumption that all demanders and suppliers are maximizers of utility and profit respectively. Beyond the peculiar pleasure some people get from such analytical exercises, not much is learned from the solution unless there are reasons given for why the equations should be true. Of course, there are reasons (viz., assumptions or equations) for why at the price being explained, the quantity demanded is at the level that it is (viz., the assumption that all demanders are maximizing their respective utility) and similarly why at that price the quantity supplied is at the level that it is (viz., the assumption that the supplier is maximizing profit) – all individuals are assumed to be optimizing and the two equations representing demand and supply are merely logical consequences of such simultaneous optimization. As discussed in Chapter 2, while it is one thing to know what the equilibrium price will be (by way of posited equations), it is quite a different thing to know how that price comes about. Traditionally, and as discussed in Chapter 7, our typical Economics 101 textbooks and teachers rely on assumptions about some vaguely specified price adjustment to correct for any discrepancy in the equation representing the equality between the market's quantities of demand and supply.[5]

Decades ago, some model builders might have been satisfied with just making such assumptions and assuming they are all true, and thereby presume to have 'closed the model', that is, to have completed the reasoning for

---

4. The remainder of this chapter uses some revised and updated versions of the methodological arguments I made in Boland [1986].

5. In Chapter 7, the price-adjustment assumptions were identified as 'Walrasian' assumptions. They are called this because such a price adjustment process is the one usually assumed to be going on in a Walrasian general equilibrium model. The economics teacher of the dialogue also brought into the discussion a 'Marshallian quantity adjustment' process, which is the one built into Marshall's theory of the supplier who responds to the difference between the current point on the marginal cost curve and the offering price of the demanders.

why the simple equation declaring that demand equals supply is true. But, as Robert Gordon has observed [1981, p. 513],

> There are . . . problems with [Arrow's adjustment equation], sometimes called the 'law of supply and demand'. . . . [T]he dynamic adjustment may take a substantial length of time. During this interval of disequilibrium, the market-clearing equality [i.e., D=S] is not satisfied and rational agents in forming their expectations about costs, prices, sales, and the rate of return on investment will not cling to the unrealistic belief that markets are continuously clearing . . .

And so, it is not difficult to see that there is nothing in such a model that tells us how long it would take for the going price to equal the equilibrium price (given the presumed equations for how the quantity demanded and the quantity supplied are respectively dependent on only the going market price). The speed of adjustment equation[6] needs to be explained such that it is seen to increase with the difference between demand and supply quantities. If in doing so it is unfortunately specified such that the going price never rises fast enough to cause the positive difference between demand and supply quantities to become zero or the price goes further to cause a negative difference, there is no guarantee that the equilibrium is reached. This means that the difference between demand and supply quantities and the speed of adjustment might both reach zero only as the amount of time necessary approaches infinity. In other words, it may easily be that the equilibrium is never reached in real time. As Hahn noted, '[I]t cannot be denied that there is something scandalous in the spectacle of so many people refining the analyses of economic states which they give no reason to suppose will ever, or have ever, come about. It probably is also dangerous' [1970, pp. 1–2]. Thus, like Hahn, some economists question the whole analytical process that so many equilibrium model builders spend their time on.

## 13.2. CLOSURE OF THE FORMAL EQUILIBRIUM MODEL

Many formal equilibrium model builders will see their task as one of assuming and specifying the additional equations and conditions required concerning the speed of adjustment like those discussed in Chapter 2 – specifically those identified by Arrow (or something that analytically serves the same purpose) such that the usual equation that claims the equality between the quantities demanded and supplied is true in real time. Arrow's version was characterized in Chapter 2 as involving two equations and one inequality: one equation for

---

6. That is specifically, equation [2.4] discussed in Chapter 2.

the speed of adjustment ([2.4]), one for assuring eventual convergence to the equilibrium ([2.6]) and the inequality ([2.5]) to assure that the price adjustment is in the right direction for convergence to market clearance. While this kind of requirement was once considered a solution to the problem of explaining the 'speed of adjustment', there are really two separate issues here – convergence and speed of adjustment – even though they are often treated as the same task. To see these issues as the same can be misleading. But before I examine this possibly troublesome issue, let us consider some of the ways in which the model of the market equilibrium is thought to have been closed.[7]

As I mentioned in Chapters 1 and 2, the classic means of closing the market equilibrium model was to assume that the market is run by an 'auctioneer' who does not allow any transactions until there is market clearance. However, I can conceive of two conceptions of the auctioneer – the 'scientist' and the 'warden'. The scientist-type auctioneer does not trust the inherent stability of the market and so, before opening the market, surveys separately the demanders and suppliers (perhaps as I illustrated with Table 2.1) and then calculates the price at which the equation representing a market clearance will be true. When the market opens, the scientist-type auctioneer just communicates the calculated equilibrium price, the transactions immediately take place and the market immediately clears.

The warden-type auctioneer starts by communicating randomly a possible price and entertains the bids from demanders or suppliers who wish to alter the price. The presumption is usually that the auctioneer does not allow transactions to take place until the bidding stops. Until this happens potential transactions are being arranged and postponed until either demand or supply runs out. But if this happens someone will need to alter the price because according to the pending transactions they are not able to maximize their profit or utility at the current price shouted out. So, such adjustments of the price is assumed to take place until everyone can accept the price. Here the warden-type auctioneer's main job is to suspend trading until such an agreement is established. While both concepts of an auctioneer are sufficient to close the model, the warden-type auctioneer is usually the one assumed.

One can easily think of many criticisms of the auctioneer approach. For example, Fisher [2003, p. 77] observes,

> It only begs the price-adjustment question to say (as is often done) that [the price adjustment] equation [[2.4]] reflects the behavior of an 'auctioneer' whose job it is to adjust prices in such a way. Most real markets do not have such specialists. Those markets that do have them are such that the specialist is rewarded

---

7. Again, being 'closed' here means only that nothing is left unexplained and thus the model can be solved for the values for the variables that the model purports to explain.

for his or her endeavors. To understand where and how such price-setting takes place requires analysis of how markets equilibrate. That cannot be done by adding equation [[2.4]] as an afterthought, nor is it likely to be done satisfactorily in the tâtonnement[8] world where only prices adjust and there are no consequences to remaining in disequilibrium.

An obvious one is that even for markets which are truly auctions these conceptions are unrealistic. Usually it is argued that the assumption of there being an auctioneer is merely ad hoc. That is, it is used solely to close the model (by establishing that the market-clearance equation [2.3] is true). Contrarily, it could be claimed the assumption actually makes the model incomplete. Consider what Fisher observes in the above quotation. If the auctioneer is necessary to run the market, one might ask whether there is a market for auctioneers and who runs that market. Perhaps, contrary to what Fisher claims in the quotation above, the auctioneer services are provided costlessly – but that would seem to require an explanation of why the auctioneer works for nothing. There is either a missing price or a missing market; either way, the explanation of why the equation for market-clearance is true is incomplete and thus the model is not closed. If model builders were to proceed without the missing market or the missing price then they would be accepting a model which violates the requirements of methodological individualism. The determination of the market price depends on the exogenous functioning of the auctioneer but the auctioneer is not a natural phenomenon. So clearly, one can say that the auctioneer is an unacceptable exogenous variable.

When the issue of closing the market-equilibrium model was a hot topic several decades ago, other ad hoc price-adjustment mechanisms were proposed. Two of the most well known are called the 'Edgeworth Process'[9] and the 'Hahn Process'.[10] About these processes, Fisher explains [2003, p. 81–83],

> Each of the two processes involves what turns out to be a deceptively simple and appealing assumption about out-of-equilibrium trade. . . . The basic assumption of the Edgeworth process is that trade takes place if and only if there exists a set of agents whose members can all increase their utilities by trading among themselves at the then ruling prices. . . . [T]he Hahn process, places a much less severe informational requirement on trades than does the Edgeworth process. In the Hahn process it is supposed that goods are traded in an organized way on 'markets'. . . . It is assumed that prospective buyers and sellers of a given good can find each other and trade if they desire to do so – indeed, in some versions . . .

---

8. Tâtonnement is the bidding and groping process that it is usually claimed to represent what Walras presumed for price adjustment.
9. See Uzawa [1962].
10. See Hahn [1960] and Hahn and Negishi [1962].

this is taken to define what is to be meant by a 'market'.... The principal assumption of the Hahn process is that markets are 'orderly', in the sense that, *after trade*, there are not both unsatisfied buyers and unsatisfied sellers of the same commodity. Only on one side of a given market are agents unable to complete their planned transactions.

The Hahn Process is claimed to be superior to the Edgeworth Process, since the Hahn Process does not *require* beneficial trades to take place whenever they are possible and the participants are not required to know of all possible beneficial trades. The Hahn Process only ensures that *after* a trade takes place all demanders or all suppliers (but not necessarily both groups) are satisfied.

I think the claimed superiority of the Hahn Process is somewhat hollow since trades are assumed to take place, yet how individuals decide to trade is not explained.[11] Furthermore, the assumptions that everyone faces the same price and that the market is 'sufficiently well organized' begs more questions than are answered. To a certain extent, these assumptions are merely the auctioneer in a disguised form. Even worse, in the Hahn Process the adequacy of the speed of adjustment is just assumed, yet it is the speed of adjustment that still needs to be explained.

These two visions of a market setting form the typical basis for specific models of explanations for why the equation specifying that demand equals supply can be true. All sorts of additional formal mathematical conditions are imposed on the postulated settings and mechanisms to prove that, under those conditions, the market-clearing equation will be true at some point in time. But while some formal equilibrium model builders of the newer culture discussed in Chapter 5 might find such puzzle solving games to be interesting, they never seem to get to the essential issue (although the issue is sometimes appreciated[12]). The essential issue is that whatever setting or mechanism is proposed, the setting or mechanism must be the result of a process of individual optimizations and not be exogenously imposed on the market.

Many other adjustment mechanisms have been proposed but none are capable of addressing the issue from a methodological individualist perspective. Why would individuals be constrained to behave as postulated in each case? Do individuals choose to behave according to the postulated adjustment process? Why do all individuals choose to behave in the same way? How would individuals ever have enough information to make such choices?

11. As was seen in Chapter 12, this is a criticism with which those theorists of the Santa Fe Institute would likely agree.

12. See Fisher [1983, ch. 9].

## 13.3. TOWARD CLOSURE THROUGH
## POSITED IGNORANCE

As has been noted often here, Arrow's 1959 article provided a suggestion for explaining how the equilibrium price might be obtained. His suggestion involves the textbooks' imperfect competitor. For his explanation one has to think of the firm as setting its price to generate a demand that just equals the profit-maximizing quantity it will produce at that price. Consider again Figure 3.3, which is reproduced here as Figure 13.1. It shows that the profit maximizing output for the given demand curve shown is $Q_1$, for which the firm will, in this case, set the price at $P_1$. This is the textbooks' view of the price-setting imperfect competitor in a long-run equilibrium.[13] Unfortunately, it has one major flaw if it is to be used as an explanation of price *dynamics* – that is, an explanation of the process of adjusting prices toward the equilibrium price.[14] Knowing the long-run curve, the firm will just jump to the one point immediately. Here, any dynamics will be in the form of the comparative statics resulting from subsequent exogenous changes in the demand curve or cost curve, rather than in the form of the endogenous behaviour of the price setter. If there is to be any endogenous adjustment dynamics, the firm must be ignorant of either the demand curve or the cost curve or both. Usually, as with Clower's ignorant monopolist of Chapter 3 and Richardson's observations discussed in Chapter 4, it is the demand curve that is in doubt since the firm is unlikely to know in advance what everyone in the market is going to demand.

How ignorant does the firm have to be so that a process of reaching the equilibrium can be explained as one of learning the details of the market's demand curve? There are many ways to deal with this.[15] It could be assumed that the firm does not know its demand curve but only has a conjecture about the demand curve's slope and position and perhaps a rule of thumb. As I suggested in Chapter 3's discussion of Clower's model of the ignorant monopolist, each time the monopolist or its agent goes to the market it uses an expected price and tries the corresponding profit-maximizing quantity and then waits to see how much was actually sold to demanders. If not all the output is bought

13. Although it would not affect the argument here, this discussion could be understood to be about a monopolist setting the price as was the case with Clower's models discussed in Chapter 3. Doing so means considering a firm operating without competition and thus there would be no guarantee that the long-run equilibrium occurs where profit is zero; instead it might be where the price (and the point on the demand curve) is above average cost at that output level.

14. For Figure 13.1 to represent the situation in the case of an imperfect competitor (unlike a monopoly), there is a degree of competitive adjustment that forces all firms to be producing where profit is zero and thereby a long-run equilibrium has been reached.

15. Some examples can be found in Robinson [1933/65]) and of course Clower [1959]. I also provided some in an unpublished 1967 textbook.

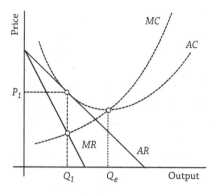

**Figure 13.1.** Imperfectly competitive long-run equilibrium (same as Figure 3.3)

or if the whole output is sold at a price higher than the trial price, either way the firm has learned just one point on the demand curve. Whenever either the quantity supplied or the expected price is wrong, it knows its expected supply quantity was not the optimum. This failure of expectations results because, as Clower put it, the firm does not know the true elasticity of demand (and hence the true marginal revenue) for its product. In effect, each trial price is a test of the demand curve's conjectured slope and position.[16]

16. The mathematics of all of this was not discussed in Chapter 3 but is worthy of taking a look at since it is rather simple and mostly the results of a few definitions. For those readers who do not have the economics jagon memorized, look again at the Prologue, note 4, for the definition of 'marginals'. With all of the following 'marginals', economists are talking about ratios of the amount of change in the cost or revenue per unit change in the level of output. When talking about a producer's situation, there is its 'marginal cost' and 'marginal revenue'. There is also its 'average revenue', which is obviously defined simply as just the total revenue ($P{\cdot}Q$) divided by the quantity sold ($Q$) – and thus average revenue is just the going price ($P$). And there is what economists such as Clower call the 'elasticity' of the demand curve (which was discussed in Chapter 2, note 15) which is merely the percentage change in the quantity demanded that would result from a one percent change in the market's price.

Now to use these definitions, assume the price has been set according to the rule derived from the necessary condition for profit maximization, namely that marginal cost ($MC$) equals marginal revenue ($MR$). By definition of $MR$, average revenue ($AR$) and the demand curve's elasticity ($e$), the following is always true:

$$MR \equiv AR\big[1 + (1/e)\big]. \tag{13.1}$$

Also by definition, $AR$ is the going market price ($P$) so when a model builder assumes that profit will be maximized for a correctly estimated $e$ (i.e. $MR = MC$), the imperfectly competitive firm's rule of thumb for setting the expected price for any given level of output will always be as follows based on what the firm thinks is the demand elasticity at the point on the demand curve corresponding to a chosen level of supply and its corresponding $MC$:

Both the imperfect competitor that Arrow would use to set the price and the monopolist whose behaviour Clower was examining in his Model III are presumed to learn by trial and error. Arrow's firm sets the price to what is presumed to be the correct price for each level of output tried. In Clower's case the firm sets the output to what the firm thinks will maximize profit. Both of their firms do this by learning to estimate the demand curve's elasticity, $e$.[17] But, unless there are very many trials it still may be the case that not much will have been learned in Arrow's case. In Clower's Model III case, the price is determined in a market rather than set by the monopolist – that is, whenever his monopolist sets its supply based on an expected elasticity and the expected elasticity turns out to be incorrect, the market price will need to adjust to clear the market for the quantity tried. In Arrow's case, though, each trial will yield additional information, but the information is useful *ceteris paribus* only if the demand curve does not shift. Still, to close such a model that purports to explain the dynamics, the model builder in both cases needs to indicate how many trials it will take to learn the true demand curve.[18] Worse, a market-based means of providing sufficient information for the convergence of the learning process only brings us back to the question of explaining the price adjustment process that clears the market whenever the firm's expectations are incorrect.

## 13.4. EXOGENOUS CONVERGENCE TO EQUILIBRIUM WITH FORCED LEARNING

While many may think the process of learning can be presumed to be inductive in situations such as this, as I noted in Chapter 12 (regarding the Santa Fe group's presumption), inductive learning can take an infinity of trials to ensure convergence; it surely would need more time than is allowed before the *ceteris paribus* assumption is no longer viable, and thus demand curves could shift. As many might see it, the real learning situation is one of estimating

$$P = MC\big[e/(1+e)\big]. \qquad [13.2]$$

Equation [13.2] is true for perfect competition as well as imperfect competition.

17. Or, as I explained in Chapter 3, the firm could use the demand curve's slope. Clower in his 1959 article referred instead to assuming something about the demand curve's elasticity. I used slope in Chapter 3 because it is easier to see and elasticity can always be calculated at any point on the demand curve if you know the slope at that point. For the mathematically inclined readers, $e$ is defined as $(\Delta Q/Q)/(\Delta P/P)$ or equivalently $(P/Q)/(\Delta P/\Delta Q)$ where $\Delta P/\Delta Q$ is simply the slope of the demand curve and where $\Delta$ is of course mathematical shorthand for 'the amount of change in'.

18. Interestingly, Arrow [1959] suggests that three would be enough.

a demand curve that is stochastically shifting.[19] Their reason might be that one could never learn fast enough to avoid the effects of shifts. Again, this is just another expression of the implicit belief that the only learning process is an inductive one. Since this belief is not usually considered a problem in contemporary equilibrium model building exercises, I will postpone further consideration of this perspective until Chapter 15. For now let us just see how it is used to close the model of equilibrium price adjustment.

Anyone who does wish to explain the price adjustment process still needs to address the difficult question: how many observations would it take to ensure that the equilibrium price will eventually be set by the imperfectly competitive price setter? If one cannot answer this, one cannot be sure that the market-clearing equation will ever be true. There are three ways in which this question is made to appear irrelevant. The first two are the Rational Expectations Hypothesis and the textbooks' long-standing implicit assumption that the market is stable with respect to both price-adjustment and quantity-adjustment behaviour. Both have already been discussed to some degree in earlier chapters. The third way is a form of argument similar to Social Darwinism that Alchian presumed in his 1950 article. In all three cases, the convergence process is exogenously given and it is merely left up to the individual to conform. Let us examine these tactics.

In Chapter 9 I offered an explanation for how the Rational Expectations Hypothesis can address this – it merely assumes that the current economic theory being used to explain the economy's behaviour is the one which has been inductively established as true.[20] The presumed inductive basis for that current theory is thus exogenous to the individuals' decision process. It is left to the individuals to use the information available to form expectations that are consistent with the current theory. When they do form consistent expectations, the economy will be in equilibrium. By assuming there is a reliable inductive learning method, one could see how individuals are forced to form such expectations when they use the same information that would be used to establish the current theory. Here, the force of inductive logic is being invoked, but no proponent of the Rational Expectations Hypothesis could have ever demonstrated that a reliable inductive logic exists.

Some readers will say that by using econometrics instead of pure induction one makes this problem exempt from the criticism of inductive learning. After

---

19. An early example can be found in Gordon and Hynes [1970, pp. 375ff].

20. In Chapter 9 the more elaborate assumption was also discussed whereby individuals forming expectations use the method of econometric estimation for which the resulting expectations are not exactly true. The main difficulty with the econometric estimation version is that it also presumes the model being used by everyone to perform the estimation is a true representation of the economy and thus all will obtain estimations that are acceptable by everyone, recognizing that it is only the observations used that are inaccurate. I will discuss econometric learning in next paragraph.

all, the most commonly asserted reason for using econometrics instead of induction is the presence of the random observation errors which econometrics was created to overcome. Instead of using observations to inductively prove one's model is true, observations are being used to confirm the conjectured model being estimated with the observations. But this just puts a confirmed model in logically the same position as an inductively proven model.[21]

Hayek was taking the same position in his 1945 article when he was, in effect, arguing for the superiority of the competitive market system over centralized planning. Unlike the pure induction version of the Rational Expectations Hypothesis, his argument did not take *successful* inductive learning as an exogenous means of assuring the convergence to an equilibrium, or of assuring that the market-clearance equation is true.[22] He implicitly assumed that all demand curves are downward sloping and all supply curves are upward sloping so that the correct information is automatically learned in the process of trial and error. But, as should be obvious by now, this argument merely assumes Arrow's additionally required equations concerning the speed of price adjustment are true as exogenous facts of nature. So, presumably, if individuals do learn when they are disappointed after going to the market, they will learn the correct direction in which to respond. And, whenever an equilibrium is reached, it will exactly be the one that is defined by the presumed stable market configuration of demand and supply curves. Moreover, *if* the individuals are ever going to learn the value of the equilibrium price, Hayek seemed to think they will be forced in this manner to learn the correct one. Unfortunately, this is much like the Edgeworth process mentioned earlier in this chapter in that it does not ensure convergence without perfect information and it does not explain how such knowledge would ever be acquired.

The third way of assuring convergence exogenously is, as I said, Alchian's way – the one I briefly discussed in Chapter 11. It is a process of reaching an equilibrium that is intended to be a lot like what many would call Darwinian evolution – that is, 'natural selection' or the 'survival of the fittest'. In economics, the fittest are the ones who have successfully made the right choice either by accident or by solving all the problems of forming expectations and maximizing in the face of uncertainties. According to Alchian's view, if the world is always limited in its resources and everything is potentially variable, one does not have to assume that each participant necessarily behaves according to the textbooks' presumption of deliberate profit and utility maximization,

---

21. This is an old story in the philosophy of science. In the 1930s it was thought that this is the best that can be done given the problems of induction. The main problem is that if all of the assumptions of an argument are not exactly true, any prediction or statement deduced from that model will not necessarily be true. And of course, that denies exactly what is needed to use a model as an explanation. For more on this see Boland [2003] and for matters of using econometrics, see Boland [2014, chs. 9, 10].

22. Neither did he presume learning could involve econometric estimation.

optimum learning processes or perfect expectations. For Alchian, such appropriate behaviour is endogenous in the sense that it is implied by the achievement of any equilibrium of *survivors*. If any firm, for example, is incurring costs that exceed its revenues, it will not survive. And eventually for the economy as a whole, since there must naturally be an equality between aggregate revenues and aggregate costs, should any one firm be making profits, some other must be making losses. Moreover, if there are profits to be had, someone will find them. So if one is considering any economy consisting only of surviving firms (and happy households) one must be looking at an economy in long-run equilibrium, that is, an equilibrium where all firms have learned enough to be making zero profits at least. And, as well, zero profits must be the best they can do.[23]

Given what most economic model builders would presumably accept as a natural fact that any economy always has a finite amount of resources, they should agree that if no one is losing money, then no one is gaining money. Thus, according to Alchian, the need to survive forces the acquisition of adequate knowledge or learning methods – at least by the survivors.[24] If all of this is extended to questions of stability, it says that Nature forces convergence regardless of how model builders explain the behaviour of individuals. But as clever as this tactic is, it still does not explain how long it would take. If there is a convergence here it is only because the convergence process is assumed to be exogenously given. This is no different than simply assuming that Arrow's market-clearance equation is true a priori and thus rendering his proposed additionally required equations concerning the speed of price adjustment unnecessary.

## 13.5. ENDOGENOUS CONVERGENCE TO EQUILIBRIUM WITH AUTONOMOUS LEARNING

A state of equilibrium is always presumed to be possible in each of these various approaches to specifying the price adjustment process in formal equilibrium models. Sometimes it is just presumed to exist in advance. But the process is always either ad hoc or exogenously imposed by circumstances. The point illustrated in the dialogue presented in Chapter 7 is that these usual ways of solving stability analysis problems may actually violate the requirements of methodological individualism. When building a complete model of the economy for which any equilibrium is stable but the stability is endogenous, the

---

23. And of course, this is exactly what one obtains in the textbooks' state of long-run equilibrium.

24. Although allowance must be made for the possibility that some just may have survived accidentally.

stability or convergence must not depend on exogenous considerations that are unacceptable for any adherence to methodological individualism. In particular, whenever a model builder successfully specifies the necessary equations but the specification is ad hoc or exogenous, the completed model forms an explanation which is either incomplete or complete only by providing by adding exogenous variables that are not natural givens. In the latter case, the resulting model is thereby violating methodological individualism.

For a long time it has been widely recognized that a minimum requirement for an equilibrium model is that any price adjustment process which fulfills the role of Arrow's additionally required equations concerning the speed of price adjustment must be derivable from the maximizing behaviour of individuals.[25] This requirement is the source of all the problems discussed in the literature concerning the disequilibrium foundations of equilibrium economics. Any shortcomings of the usual attempts to specify equilibrium models are almost always due to failures to recognize this requirement. To understand this requirement, its implied procedural rules for the equilibrium model builder need to be examined.

The singular behavioural assumption of the utility (or profit) maximizing individual has always been paradigmatic in neoclassical explanations. It is not clear whether the neoclassical paradigm can ever adequately represent all aspects of the problem of constructing an optimal price adjustment mechanism. The speed of adjustment is not usually considered a direct source of utility; that is, it is not desired for its own sake. The price-adjustment speed is merely a means to the acquisition of final goods from which the utility is derived. Few people drink wine (or beer) for its own sake but do so for its alcohol content, among other attributes. The sources of the utility are the various attributes (or 'characteristics').[26] So, should formal model builders view the price-adjustment speed in a manner whereby it is just a characteristic, they are not putting it beyond the domain of choice theory.[27] All that is required is a mathematically representable mechanism that shows how the price-adjustment speed affects the quantities of final goods. This mechanism is not apparent in models built using assumptions such as those found with the Hahn Process. Nevertheless, the specification of such a formal mechanism

25. For example, see Gordon [1981, p. 512] and Fisher [1981, p. 279].

26. See Kelvin Lancaster's 1966 article which provides a model of consumers behaviour where they buy products for their characteristics such as, say, calorie content or fiber content. In his model, preference maps are about choosing characteristics rather than commodities. Commodities for Lancaster are just the means to the end, not the end itself.

27. For example, if the model builder does recognize time in the adjustment process and thus has individuals seeing time lost in the process as a cost in terms of a disutility, then the speed of adjustment would matter to decision makers. Gary Becker raised just this issue of a disutility of misallocated time in his famous 1965 article, 'The theory of the allocation of time'.

seems to be the ultimate purpose of the equilibrium model builders interested in stability analysis – and it is not totally unreasonable that such a mechanism might be constructed.

It must now be asked, will any such mechanism do? Or are there limits on what can be assumed in the process of constructing such a mechanism? Apart from satisfying the formal requirements for optimizing or for obtaining an equilibrium in a model according to mathematical standards and techniques, there are really only the requirements of methodological individualism that I discussed in Chapters 6 and 7. If the mechanism is to be consistent with neoclassical theory, any alleged exogenous variable which is non-natural and non-individualist will need further explanation by acceptable means. A typical example of this requirement occurs in the explanation of the price-adjustment mechanism using monopoly theory as was the case in Clower's 1959 article. For a monopoly to exist – or for that matter, anything less than perfect competition – there must be something restricting competition. Is that restriction exogenous or endogenous?

Few if any of the well-known imperfect-competition stability models provide an explanation for *why* there is less-than-perfect competition. But, whenever any complete explanation is consistent with the psychologistic version of methodological individualism,[28] a long-run equilibrium model of price-takers is assumed. Given that psychologism[29] is almost always taken for granted in neoclassical economics (since the individual is always identified with his or her exogenously given utility function), one wonders whether explanations of stability based on imperfect competition will ever satisfy all honest neoclassical equilibrium model builders.

28. This version of methodological individualism will be explained in Chapter 15. For an earlier discussion, see Boland [2003, ch. 2], which is about all of the various versions of methodological individualism incorporated in mainstream economics.

29. Psychologism is the methodological presumption that all explanations of all social events must be reducible to psychology-based explanations.

# CHAPTER 14

✦

# Building models of non-clearing markets

The idea of a non-clearing market is not central in the discussions of neo-classical microeconomic models. However, when it comes to discussions of old Keynesian macroeconomic models, so-called persistent unemployment (as a form of a non-clearing market) once fostered an extensive literature and debate about disequilibrium models. As Kaldor [1983, p. 13] explained:

> [M]ost of the debate around the legitimacy of Keynes' notion of 'underemployment equilibrium' was misplaced. It is the notion of a 'full employment equilibrium' which is an artificial creation, the consequence of the artificial assumption of constant re-turns to scale in all industries and over the whole range of outputs which implies infinite divisibility of everything ... [M]ost of the voluminous literature concern-ing the reconciliation of Keynesian analysis with Walrasian general equilibrium – in terms of 'disequilibrium' economics, inverted velocities of price and quantity adjust-ments, absence of the 'heavenly auctioneer', etc. – is beside the point.

The object of most of these Keynesian disequilibrium models was to explain how an 'underemployment equilibrium' (as Kaldor calls it in the quotation above) is possible even when everyone is assumed to be an optimizer. Traditionally, the per-sistence of such unemployment is explained as being the result of 'wage rigidities' but there is seldom any reason given for why the price of labour is not flexible.[1]

---

1. I think it is interesting that in the mid 1970s persistent unemployment was occurring at the same time as a high rate of inflation. Such a coincidence was given the label 'stagflation' and was deemed by some neoclassical model builders to be im-possible to explain with a Keynesian model. But note that it is explainable if one

Equilibrium models designed to explain *persistent* unemployment are unlikely to be accepted by ordinary neoclassical macroeconomic equilibrium model builders – particularly if they involve exogenously fixed prices, since an unexplained fixed price would violate the requirements of methodological individualism. Furthermore, even some Keynesians do not accept such a definition of Keynesian economics. One does not have to think that Keynes was arguing in favour of a fixed-price explanation of unemployment – all that might be required is that the wage-rate's speed of adjustment be limited so that it is not as fast as the price adjustments for final goods. Nevertheless, one can readily understand the neoclassical rejection of Keynesian macroeconomic models whenever the existence of a fixed price is used to distinguish Keynesian from neoclassical models.

While persistent unemployment will be one of the topics of concern in this chapter, the discussion will not be limited to that topic. This chapter will also discuss explanations of a state of a disequilibrium, for which the usual calculus properties of an equilibrium are not appropriate. And it will look at an alternative view of Keynes' so-called macroeconomics to see whether he may have already introduced the means by which alleged methodological limitations of neoclassical equilibrium economics can be avoided.

## 14.1. UNINTENTIONAL DISEQUILIBRIA

One might wonder whether any acceptable neoclassical model could ever explain the persistence of a disequilibrium.[2] As Jean-Pascal Benassy saw this, 'Most concepts of conventional (or neo-classical) economics hold rigorously only in general equilibrium, which precludes the study of Keynesian or Marxian Economics, or a satisfactory integration with macro-economic theory since all of these are essentially concerned with disequilibrium states, where transactions take place at non-Walrasian prices' [1975, p. 503]. Of course, the appearance of a disequilibrium is easy enough to explain away. We could just say that it is only a temporary phenomenon which disappears once we broaden our perspective by asking whether the disequilibrium would

gives up explaining the coincidence using comparative static methodology, which always starts with a state of long-run equilibrium and then changes just one exogenous variable to create stagflation. If one instead builds a model in long-run disequilibrium (which is possible with how Keynes would see the matter), stagflation is easy to explain.

For an extensive discussion of macroeconomic disequilibrium theory, see the 1980 article by Allan Drazen.

2. The some of the remainder of this chapter includes revised versions of the methodological arguments I made in Boland [1986].

persist in any long-run situation.[3] The task at hand is to consider how a state of disequilibrium such as 'involuntary unemployment' could persist for a significant amount of time and could still be explainable in terms consistent with methodological individualism.

Any problems concerning the explanation of persistent unemployment are relevant for all explanations of non-clearing markets. Since the usual reasons given for any market's failure to clear are, as I noted about persistent unemployment, either that the price is being held rigid or that it does not change fast enough, such an explanation only begs the question of why the price is rigid or relatively inflexible. So, when the disequilibrium model builder turns to explain why prices are rigid, what are the usual exogenous variables considered? And, will their exogeneity violate the requirements of methodological individualism?[4]

Perhaps one might try to identify some particular variable to be exogenous in order to explain why the prices fail to adjust fast enough – but if one does this, all that would usually be created is a model with a so-called temporary equilibrium which merely plays the same role as Marshall's short-run equilibrium. In a temporary equilibrium either the price or the quantity is held fixed while the other variables are allowed to be the only means of adjustment. The question for models of this Marshallian type is whether it makes sense at all to discuss equilibria when one is trying to explain disequilibria.[5] This question arises for those models which try to base the rigidity of the price on the textbooks' imperfectly competitive market structure.[6]

The textbooks' imperfectly competitive equilibrium model usually implies a certain kind of disequilibrium in the sense that such an equilibrium

3. Doing this merely avoids the challenge and so such a tactic will be of no interest here. The tactic of claiming that any observed unemployment is really voluntary will also be avoided as it, too, is just a means of avoiding the challenge.

4. When it is thought that the adjustment of prices must be explained by introducing appropriate implications of imperfect competition models, the question is then begged as to why there is a barrier to entry into the industry. One might wish to explain the choice of market structure so as to render it endogenous. But, what new exogenous variables are introduced in this step? Usually, it is some sort of exogenous transactions cost schedule. For examples, see Coase [1937] and Williamson [2000]; see also Loasby [1976]. This approach begs the question of what exogenous variables determine the transaction cost. If the transaction cost is in any way influenced by prices, the explanation becomes at best incomplete or at worst circular.

5. Moreover, this type of explanation of a disequilibrium is merely about a process that will make sense only when the eventual equilibrium is reached.

6. Presuming imperfect competition to explain the existence of a non-cleared market is a mixed blessing. Under certain interpretations (see Boland [1986, Chapter 2]), the explanation merely presumes another type of equilibrium, and thereby precludes the possibility of disequilibrium. Under other interpretations, such as comparisons with ideal states of perfect competition, the explanation of rigidity implies some sort of sub-optimality and hence implies that at least one market is not in equilibrium.

will involve an amount of excess capacity[7] – but any excess of capacity amounts to a non-optimal use of resources when compared to a perfectly competitive equilibrium. And, as already mentioned, there is still a question of why there should be such an imperfectly competitive market structure.

The textbooks' explanation of the imperfectly competitive firm conveniently gives room to recognize that the knowledge requirements for an imperfectly competitive equilibrium are always much more demanding than those of a perfectly competitive equilibrium with the latter's equilibrium-price takers. And this is the primary reason why many theorists such as Arrow turn to the textbooks' imperfectly competitive firm to either explain price dynamics as I discussed in Chapters 2 and 3 or explain the non-clearing market. But, to guarantee that the textbook's imperfectly competitive equilibrium will be reached, the firm which is not a price-taker must know the whole demand curve it faces. As I discussed in Chapter 4, knowing the whole demand curve before putting one's product on the market necessitates knowledge of what every consumer is going to demand at every conceivable price. And of course, I subsequently critically discussed the matter of model builders who think this knowledge can be acquired inductively. Obviously, knowledge of the whole demand curve requires too much for any realistic imperfectly competitive equilibrium, but that is all right since, for many equilibrium model builders, it seems to provide an essential reason for why, at any one point in time, there might be a disequilibrium.[8] Any disequilibrium (in terms of universal maximization) is easily explainable as the failure of demanders or suppliers to optimize due to misperceptions of the relevant constraints, and this is exactly what was discussed in Chapter 3 with Clower's model of the ignorant monopolist that ends with the firm not truly maximizing. It nevertheless can be pointed out that although for Clower's Model III the market ends in what appears to be an equilibrium in the sense that the firm sees no reason to change, it is not the equilibrium discussed in textbooks, for which an equilibrium would always assure universal maximization by all the participants in the market.

---

7. What economists call excess capacity is perfectly illustrated in Figure 13.1. In that figure the firm is shown in a long-run equilibrium at point $Q_1$. The firm is in such an equilibrium because at that point its profit is both maximized ($MR=MC$) and zero ($AR=AC$) but at that point the firm also has the capacity to increase output and lower average cost. The possibility of lowering average costs means that less of one or more of society's resources can be used per unit of output. For more about excess capacity in imperfect competition, see Boland [1992, ch. 5].

8. For example, see Fisher [1983, p. 190].

## 14.2. ENDOGENOUSLY DELIBERATE DISEQUILIBRIA: KEYNES-HICKS GENERALIZED LIQUIDITY

Any disequilibrium in a neoclassical equilibrium model always implies that someone is failing to maximize short-run utility or profit. For the labour market, an unemployment (i.e., excess supply) equilibrium means that some workers are willing and capable of providing more labour than is demanded. Excess demand for labour would mean that some firms are using less labour than they desire and thereby are producing less than their capabilities allow. In a product market, a disequilibrium means either some consumers are being forced (because of excess demand) to purchase less than their budget would allow or that some suppliers (as a result of excess supply) are selling an amount of its output that is less than what would be its profit maximizing level of output. In other words, a disequilibrium failure to meet one's objective in the market is always seen as one of being somehow forced to choose a point that is not optimal. Is the reverse true? That is, whenever people are seen to be operating inside their capabilities, must this be evidence of a disequilibrium?

Keynes, in effect, asked model builders why they think that the individuals who are not operating on the boundaries of their capabilities are actually failing. This question is inherent in his assault on what is now called neoclassical economics – particularly on its reliance on long-run equilibrium states. It is a question which puts all the concern over disequilibrium model building into an entirely different light. According to Keynes [1937, p. 223]:

> I doubt if many modern economists really accept Say's Law[9] that supply creates its own demand. But they have not been aware that they were tacitly assuming it. Thus the psychological law underlying the Multiplier has escaped notice. It has not been observed that the amount of consumption-goods which it pays entrepreneurs to produce is a function of the amount of investment-goods which it pays them to produce. The explanation is to be found, I suppose, in the tacit assumption that every individual spends the whole of his income either on consumption or on buying, directly or indirectly, newly produced capital goods. But, here again, whilst the older economists expressly believed this, I doubt if many contemporary economists really do believe it. They have discarded these older ideas without becoming aware of the consequences.

9. Many neoclassical economists think this 'Law' holds presumably because to supply a quantity of some commodity, the producer had to be paying out to somebody the market value of that quantity in terms of wages, capital costs, interests and profits. Thus, there is necessarily enough money available to buy what is produced. However, it could be argued that just because the money is available to buy what has been produced, this does not mean there is a demand for it unless the producer has perfect knowledge of the market's demand.

For those readers who might not remember their introductory macroeconomics class[10] and thus do not understand what Keynes was talking about in
the above quotation, let me review the consequences that he had in mind.[11]
What is still not appreciated is the contradiction between what Keynes
called the 'psychological law underlying the Multiplier' and the neoclassical microeconomics method of explaining the consumer. The 'psychological
law' he is referring to here is simply the idea of an *exogenously* given psychological 'marginal propensity to consume' that was discussed in Chapter 5. As
students might learn in some elementary macroeconomics classes, students
are to assume that an individual never spends all of an extra dollar of income
earned but just some fraction of it. That Keynes would take this 'law' as a
psychological given might cause some concern, as it is not directly related
to the microeconomics textbooks' idea of a utility-maximizing consumer
facing a given income or budget. The microeconomics textbooks' consumer
is usually thought to spend all of his or her budget. If the income of the
microeconomics consumer increases, planned purchases will be expanded to
fully spend the extra income so as to be on the boundary of the consumer's
capabilities, that is, to be operating on the boundary of his or her income
constraint.[12]

In distinction from what Keynesian models are presuming, it is important to remember here that neoclassical equilibrium methods of explanation
always see all individuals operating on the boundaries of their capabilities.
Of course it is easy to avoid this by recognizing *savings* as a variable of choice
such that the budget would then be less than the income. I suspect that most
microeconomics textbooks do not do this since it makes explaining the theory
of the consumer more complicated. Those that do address saving simply have
savings as an object of choice even though this would presume perfect knowledge of the future in order to be making an optimal choice as is done with
respect to commodities such as bread or butter.

10. Or have taken a macroeconomics class where the teacher or textbook carefully
avoided talking about Keynes' macroeconomics.

11. And particularly examine those aspects which differ from what one would
learn in a typical microeconomics class.

12. In microeconomics, little is said about the difference between income and
budget since the issue at hand is usually only about choosing what or how much to
consume of commodities. The idea of people saving rather than spending all of their
income is usually discussed only in the macroeconomics class rather than in the microeconomics class. One reason for this is that the interest rate is rarely discussed
in the microeconomic textbooks. When it is discussed, it would be about deciding
between present and future consumption – waiting to consume in the future one
might be seen to put part of the income in the bank and earn interest instead of
consuming today.

### 14.2.1. A macroeconomics textbook's simple Keynesian macroeconomics equilibrium model

In order to discuss endogenously deliberate disequilibria, I need to review the usual textbooks' simple macroeconomics equilibrium model that I briefly discussed in Section 5.5 of Chapter 5. Macroeconomics models are merely models of aggregated microeconomic variables. If one were to characterize a Keynesian view of the microeconomics' consumer, it would be something like $C_i = \alpha + \beta Y_i$ for consumer $i$. The $C_i$ represents the dollar value of the total amount of consumption goods that consumer purchased given his or her income $Y_i$ that forms the boundary for consumption. And, whenever there is no savings, the $\alpha = 0$ and the $\beta = 1$. When one aggregates all of the economy's individual's consumption to form the total national demand for consumption goods ($C$) given the aggregated total national income of all consumers ($Y$), one, of course, gets the following Keynesian equation:

$$C = \alpha + \beta Y_s \qquad\qquad [14.1]$$

The variable $Y_s$ represents the aggregate total national income (GDP) and is the result of payouts from producing the national income.

For Keynes, what one would call savings, he calls not-consuming. This is because he allows for having the consumers choose to hold money back from consumption as a matter of providing 'liquidity' (I will discuss this in the next sub-section). In a simple Keynesian macroeconomics equilibrium model, the aggregate total demand $Y_D$ for *all* goods – consumption demand ($C$) plus the exogenous[13] demand for non-consumption goods ($I$) – which can include the sum of both exogenous business investment and exogenous government investment – is simply expressed as:

$$Y_D = C + I \qquad\qquad [14.2]$$

And for this to be an equilibrium model, the total demand, of course, must equal total supply:

$$Y_D = Y_S \qquad\qquad [14.3]$$

Equation [14.1] is where Keynes' psychological law appears, allowing for non-consumption (savings plus liquidity), and hence the $\beta$ is claimed to be a

---

13. This is because this is the simplest version of the model. As I said in Chapter 5, more elaborate models might provide an explanation of the demand for investment goods (e.g., productive capital).

positive fraction less than 1. While β being a positive fraction less than 1 is required for equilibrium stability (as is the requirement that also α be positive), for Keynes it is just considered a psychological given.[14]

Here unlike what I began with in this sub-section, the distinction between micro- and macroeconomic definitions of the variables in this simple macroeconomic equilibrium model is usually not mentioned in macroeconomics textbooks. Of course, the macroeconomic variables are all aggregates, except for the psychologically given parameter β.[15] It does not matter whether all individuals have the same β so long as Keynes' psychological law is true in aggregate. Whenever individuals differ regarding their personal marginal propensities to consume, the β in equation [14.1] is merely something like the average for all consumers.

## 14.2.2. Choosing not to consume in simple Keynesian macroeconomics equilibrium models

According to Keynes, it is important to recognize that individuals do not operate on the boundary of their individual capabilities. One could successfully operate on one's boundary only if one were absolutely certain about the future. Given uncertainty, it might be wise to leave a little room for error or for the

---

14. Such a simple equilibrium model would usually be used in textbooks to determine the effect a marginal change in the one exogenous variable, *I*, would have on the equilibrium solution employing the Marshallian comparative static analysis method that was discussed in Chapter 1. Using his method one would calculate the equilibrium aggregate national output and national income, that is, for when equation [14.3] is true. Let that equilibrium aggregate amount be labeled *Y* and then calculate the derivative: *dY/dI*. This derivative is called the investment multiplier and would be determined as follows using the current value of β:

$$dY / dI = 1/(1-\beta).$$

This derivative is called an *investment* multiplier but it could also include or represent a result of a change in a government's investment – perhaps in more infrastructure. In this case (or when government expenditures are recognized separately in a more complicated model), the multiplier refers to the notion that if a government increases its expenditures by, say, 10%, the effect that has on the GDP (*Y*) will be a multiple of that increase.

As a matter of simple algebra, Keynes' psychological law (viz., $0 < \beta < 1$) is essential for having an investment multiplier greater than one (as Keynesian proponents of government stimulation would presume), as well as for ensuring that an equilibrium *Y* exists. What is most important to recognize here is that his 'law' requires all individuals to be operating inside their income constraints, or simply that they are not spending all of any *extra* income they receive.

15. The parameter α is usually not explained but it might be seen as the minimum income necessary for survival.

unexpected. Many people save for this very reason and not just to earn interest on their savings. Putting one's savings in the bank just to earn interest would be another form of optimization and this is not what Keynes meant by saving. As I said, for Keynes, saving is simply 'not-consuming'. For him, there are other reasons for not consuming and for β being less than one.

Keynes' famous 1936 book is, for the most part, about the consequences of this contradiction – specifically the contradiction between what Keynes called the 'psychological law underlying the Multiplier' and the neoclassical microeconomics method of explaining the utility-maximizing consumer, used by those economists who wish to continue using the Marshallian-type neoclassical equilibrium method of explanation. Despite what Keynes later said in a 1937 article, in the 1950s and 1960s when Keynes' *General Theory* was thought to be the right basis for building macroeconomic models, most students were taught that the significant aspect of his book was his emphasis on both 'expectations' and 'liquidity'. Unfortunately, most students were taught that Keynes' 'liquidity' was only important for his considerations of monetary policy effectiveness.[16] This missed the major point of his criticism of neoclassical equilibrium economics. For Keynes, the savings one puts in the bank may only be part of one's chosen liquidity. The essential importance of liquidity is that it can also represent a deliberate choice to be inside the boundary of one's capabilities and thus represents a direct conflict with neoclassical methodology at a fundamental level.

It is not easy to see this aspect of the idea of liquidity in Keynes' book because he is usually seen to be presenting it primarily in terms of financial liquidity. Of course, financial liquidity is closely related to the question of investment that concerned Keynes. What needs to be seen is how important the concept of liquidity is for understanding what Keynes meant by 'the consequences' (see the end of the above long quotation from page 223 of his 1937 article). But to see this, a more general concept is needed. Hicks years later actually provided such a general view in his 1979 book [pp. 94–95]:

> Liquidity is freedom. When a firm takes action that diminishes its liquidity, it diminishes its freedom; for it exposes itself to the risk that it will have diminished, or retarded, its ability to respond to future opportunities. This applies both within the financial sphere and outside. I have myself become convinced that it is outside the financial sphere (very inadequately considered, in relation to liquidity, by Keynes) that liquidity is potentially of the greater importance. . . .

16. Specifically, Keynes said that when interest rates are low, the incentive to put one's savings in the bank is also low. As a result, governments that increase the money supply (by buying outstanding government bonds) will not find much of a benefit as people will just put the monetary proceeds in their pockets. Keynes called this governmental situation the 'Liquidity Trap'.

Liquidity preference, for the financial firm, is a matter of marginal adjustments, as Keynes very rightly saw. But the liquidity problem of the non-financial firm is not, as a rule, a matter of marginal adjustments.

For the Marshallian world of comparative statics[17] Hicks is arguing that there is always enough time to make marginal adjustments and thus there is no need for liquidity. In the real world where many things are happening simultaneously, the Marshallian method of explanation is usually misleading. The keystone of Hicks' argument is the idea that every decision maker forms a 'plan' based on the perceived givens, constraints and prices. Whenever decisions take time to execute, the passage of time means that the plan's original givens might have changed – or may even have been wrongly perceived. By the time the decision plan's execution is completed the resulting decision may no longer be optimal.

Automobile manufacturers, for example, might think that the future will always favour large fuel-inefficient personal automobiles. If they also think there is an unlimited amount of inexpensive fuel, their optimal plan might be to specialize and invest in the production and marketing of such autos. If the market should abruptly shift in favour of more efficient automobiles – perhaps because of an increased availability of electric or hydrogen powered autos or just more small fuel-efficient autos – or if the supply of inexpensive fuel disappears perhaps as the result of a fuel tax, the profit potential for the firms producing inefficient automobiles will be drastically altered. Such examples might be too dramatic for ordinary decision making, but the same possibility would exist whenever a specific size of a market is anticipated by one firm, but subsequently there is a sudden increase in its demand due to a strike or fire at a competing firm. In either case, if the previous level of planned output was the one corresponding to the usual neoclassical or Marshallian long-run equilibrium (i.e. the output was set to where price equals average as well as marginal costs), there is no extra capacity since it would not have been needed. Here, the firm would not be able to quickly respond competitively to the new market potential by increasing output by much (even though the price may have risen above average cost[18]). It could respond only if the firm

17. It is important to always keep in mind that 'comparative statics' refers to a method of using an equilibrium model. As I explained in Chapter 1, one would change just one of an equilibrium model's exogenous variables and then see how that changes the equilibrium solution to that model. One then compares the new solution to the original solution that would have been obtained before changing the exogenous variable.

18. This is presuming that the marginal cost curve rises steeply. Real firms do not behave like those in our microeconomics textbooks. It is more likely that a firm produces as much as its acquired capacity would allow and so expanding output would either be impossible or one more unit of output would be much more expensive to produce than a small increase in the price would allow.

was not actually producing on the boundary of its production capabilities, but this would be contrary to the requirements of a neoclassical long-run equilibrium. An increase in capacity would take time but, as always, even if the firm immediately invests to increase capacity, by the time the new higher capacity is realized it might not be an optimum capacity. The conditions that prompted the capacity increase, such as the strike or the fuel shortage, may be over. It would seem that zero excess capacity – that is, the absence of any liquidity in the non-financial sense – would be sub-optimal. However, as discussed in Chapter 4 with Richardson's warnings that in a changing world, a true optimum with respect to excess capacity or liquidity may not be knowable by the firm because its calculation depends on the unknown contemporaneous happenings and decisions of other people. And so, calculations are made even more difficult whenever their optimality depends on the unknowable future. And as Hicks was saying, this is exactly why Keynes saw a need to recognize liquidity.

Hicks' argument here is that, fundamentally, liquidity is flexibility and as such it is a deliberate choice variable. Moreover, from the viewpoint of Keynes 1937 article, such flexibility is simply good business practice (as illustrated in the examples I discussed above). It is not, however, just about investment. Whenever the labour market is not clearing because the current real wage is above the one which would clear the market, there is excess supply and thus by neoclassical standards, there is a sub-optimal disequilibrium. But, from the Keynes-Hicks viewpoint, such excess supply may very well represent a desirable state for the employer. For some firms the ability to expand production immediately whenever necessary is a desirable position.[19] This may also be true for the employee. A thirty-five hour work week can be an optimum for an individual, even though he or she is capable of working a fifty hour week at the going wage-rate. Leaving a little free time for picking up emergency money when it is needed may be more desirable than signing a contract to work to one's limits and thereby precluding any leisure time.

This is not just about questions of static capabilities. It may be desirable to have the ability to choose one's speed of adjustment to changing conditions. Sometimes, a fast response is more appropriate than a slow response and at other times it is the reverse. Flexibility is the key idea here. But is flexibility a variable that can be chosen in the same way one would choose a quantity of food or a quantity of capital required to achieve the current objective? Both Keynes and Hicks seem to be arguing that one's choice of liquidity, be it financial as Keynes discussed or non-financial as Hicks noted, is not a variable that is amenable to Marshallian optimization analysis. The type of flexibility

19. Interestingly, unlike the time that Keynes and to a certain extent Hicks was writing, for some firms today this is made very difficult because of the 'just-in-time' parts supply method used to minimize inventory costs.

or liquidity that is appropriate for any conceivable situation always depends on the future value of variables that cannot easily be determined. However, knowledge of the variables affecting the choice of an optimum plan would be essential for the usual neoclassical explanation even when those variables are thought to be merely stochastic distributions.

In the typical neoclassical textbook equilibrium theory of the firm or individual, liquidity is not usually a variable being considered. To appreciate the significance of stressing the desirability of liquidity one needs to see why it is not part of the usual neoclassical equilibrium model. Consider Figure 14.1, which merely represents a firm's production function for good $Y$ using the available labour input $N$. What the production function really shows is the *physical maximum* amount of $Y$ that can be produced at each level of input – that is, it shows the firm's productive capabilities. The production function is the boundary of productive capabilities. But, it is the shape of this boundary (i.e. its slope $dY/dN$) that is used to determine the optimum combination of input and output levels. The usual textbook assumes that the firm, of course, chooses the combination which maximizes the difference between revenue and cost for the given prices. The total real cost illustrated in Figure 14.1 is the sum of the fixed real cost, $A$, plus the variable real cost, $N \cdot W/P$, where 'real cost' again means measuring the quantity of input in units of the product produced (e.g. multiplying the short-run labour input by its price and dividing by the price of the produced good). Given that it is assumed that the firm faces equilibrium prices for both the product and the inputs, so long as those prices do not change, the optimum slope of the firm's total cost curve is thus a constant $W/P$ for the firm facing an output price $P$ and a labour input price $W$. Thus as the neoclassical textbooks will say, to be maximizing profit, the firm will choose the short-run labour input-output combination for which:

$$dY/dN = W/P. \qquad\qquad [14.4]$$

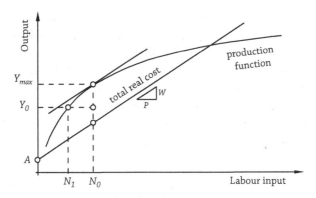

**Figure 14.1.** Optimal input-output vs. liquidity allowance

So far this is fairly elementary as illustrated in Figure 14.1 – the neoclassical optimum point will thus occur in the usual neoclassical way where, at the chosen level of labour input $N_0$, the slope of the production function and the slope of the total real-cost curve are equal. And this is the essential point of this elementary discussion. However, when the firm chooses to allow for some non-financial liquidity, it in effect chooses to be below the boundary formed by its production function (such as output level $Y_0$ for input level $N_0$). In doing so, the slope of the production function cannot be used to explain the firm's choice of an input-output combination since one cannot be sure whether this is a decision to waste input $(N_0 - N_1)$ or to stay below the maximum output level, $Y_{max}$.

The price system for a perfectly competitive market requires that whenever prices or price *changes* are to matter, it is essential for all neoclassical firms to be operating on the boundary of their capabilities such as $Y_{max}$. In such a perfectly competitive case, if either $W$ or $P$ were to change in my illustration there would be a predictable reaction along the production boundary.[20] While it is not shown in my illustration, being on the boundary is essential for all the arguments in favour of the ability of a competitive price system to produce a socially desired optimal allocation of resources in the long run. Such optimality does obviously require that inputs are not being wasted. Some might see that being on the boundary is a minimum requirement for efficient production.

That everyone should use prices as appropriate information in making decisions about what to produce or buy is the essential idea for those who promote the neoclassical competitive market system. When the price of fuel-efficient autos is rising relative to inefficient autos, such a price increase is important social information for the auto producing firms. If those firms respond to such a price increase by increasing the output of efficient autos, the firms are doing just what society wants.[21] But what happens to the competitive market system when the firm is not operating on its capabilities boundary – that is, for example, when it is deliberately providing liquidity or flexibility in the form of excess capacity? For one thing, equation [14.4] may not be satisfied[22] and so excess capacity means profit is not necessarily being maximized with respect to the available level of the input. Worse than this, the market's prices no longer act as appropriate information for other decision makers. The competitive market system in this case will not necessarily lead to the 'best of all possible worlds'.

20. For example, if the price P were to increase in Figure 14.1, the slope of the cost line would be less steep and the optimum point would move to the right, meaning that more input is used to produce more output which is what was desired by the market as indicated by the increased price.

21. Although this need not be intentional, of course.

22. I illustrated such a situation in Figure 14.1. For the purpose of providing liquidity the firm may have hired $N_0$ but uses something less, between $N_0$ and $N_1$.

Keynes stressed the recognition of liquidity as a matter of precaution in any decision plans which take time to be executed. Liquidity or flexibility may also be seen to be a means the firm can use simply to avoid the difficult task of calculating the optimum decision plan whenever it is recognized that ignorance of the future is likely. And there is always the logical possibility that an optimal amount of liquidity or flexibility has been chosen accidentally. It is important also to recognize that any claim that a firm is not optimizing does not deny a conscious attempt on the firm's part to choose an optimum amount of liquidity or flexibility. But, of course, given any ignorance about the future it would be unlikely for the firm to always be successful in such a choice. However, even if a model builder thought it was possible for a firm to choose an optimum amount of liquidity or flexibility, there would remain one overwhelming problem: the idea of an optimal amount of liquidity or flexibility is self-contradictory. If liquidity or flexibility could be chosen just as any other productive input, there would be no need for liquidity or flexibility.[23] So, it is quite possible that whenever a necessary role for liquidity or flexibility is recognized, one is thereby also recognizing what amounts to a deliberately chosen disequilibrium relative to the equilibrium defined in the ordinary neoclassical equilibrium model's explanation of demand or supply.

A choice variable like the Keynes-Hicks concept of general liquidity may immediately explain the existence of a persistent excess-supply disequilibrium. For such an explanation one must continue to define the supply curve as that indicating the supply that would be chosen according to a neoclassical optimization explanation. If the firm were producing to its full capacity as might be required by a maximization process, the supply would be greater than what is supplied when a provision is made for a certain margin of non-financial liquidity. This viewpoint, of course, merely raises the question of whether there is an optimum amount of liquidity. If such an amount of liquidity could be defined, liquidity would be just another choice variable like capital itself; and so there would be no undesirable persistent excess supply since the amount supplied was the optimum output. Obviously, optimal liquidity is not what either Keynes or Hicks had in mind.

## 14.3. DELIBERATE DISEQUILIBRIA
## VS. METHODOLOGICAL INDIVIDUALISM

Anyone attempting to provide a methodological-individualist explanation of the persistence of a disequilibrium faces an interesting dilemma. Obviously, a disequilibrium can be considered either unintentional or intentional. The

---

23. See further, Boland [1992, ch. 9].

choice, however, is not arbitrary. When the disequilibrium is explained as an unintentional consequence of intervening exogenous variables, those variables would have to be explained if they are neither individualist nor natural givens. But, once one explains those exogenous variables, one would have in effect explained the disequilibrium away. This, of course, is a violation of the original task, which was to explain the persistence of the state of disequilibrium rather than explain why it does not exist.

I think it is doubtful whether there could ever be an acceptable neoclassical explanation of a persistent disequilibrium. Every neoclassical explanation must view the disequilibrium as being the consequence of the intentional acts of autonomous individuals. In this regard, the Keynes-Hicks concept of deliberate liquidity or deliberate flexibility is a denial of deliberate short-run optimization but it would seem to hold more promise of an internally consistent explanation of disequilibria than would the neoclassical concept of deliberate maximization with respect to liquidity or flexibility. And more important, the choice of some liquidity or flexibility instead of just optimization is clearly an act of autonomous choice. By being inside one's limits, one is not forced to make choices that are uniquely defined by circumstances, as would seem to be the case in so many neoclassical equilibrium models.[24]

The term 'autonomous' is being used here because I wish to stress that the individual does not have to be identified with his or her psychological state, as is commonly done in neoclassical equilibrium economics. But I am also stressing this because distinguishing between 'autonomous' choices and psychologically determined choices (e.g. $0 < \beta < 1$) highlights an important aspect of Keynes' criticism of neoclassical equilibrium models. Like most neoclassical economists, Keynes obviously seems to have accepted psychologism[25] with his identification of individuals with their psychological states. Nevertheless, the deliberate use of liquidity, whether it be in the form of excess capacity or the marginal propensity to consume a fraction of any extra dollar of income, still directly confronts the neoclassical presumption that individuals are optimizing and thus operating on the boundaries of their capabilities (i.e. where $\alpha = 0$ and $\beta = 1$).

If I have made my case it should be evident that any neoclassical explanation of disequilibria as intentional states of affairs is necessarily self-contradictory; but also that such is not the case for the Keynes-Hicks explanation based on deliberate liquidity or deliberate flexibility. Nevertheless, some may still question whether a Keynes-Hicks explanation could ever be both complete and consistent with the requirements of methodological individualism.

24. About this, see the interesting methodological discussion in Latsis [1972].
25. It could, of course, be recognized that if Keynes wanted neoclassical economists to listen to his criticisms, putting it their terms might make that easier and thereby help.

# CHAPTER 15

✿

# Building models of learning
# and the equilibrium process

Throughout Part II the discussion was about questions concerning stability or disequilibria and during that discussion I repeatedly raised questions involving information and learning, questions that are still pending
today. In this chapter I will be laying out what I hope has been learned about
building equilibrium models in which learning and knowledge does matter.
Disequilibrium theorists in the 1980s appreciated that a price-adjustment
mechanism must include a role for learning and information, but they all
seem to have presumed that there is only one way to learn.[1] There are two
problems with this presumption. One concerns why it is presumed that there
is only one type of learning method to consider,[2] and the other concerns the
nature of the one presumed type of learning method. Consider some examples
such as Davidson: 'Neoclassical microanalysis provides a pure theory – completely deductive – to show the existence of a long-run, static full employment equilibrium, while Keynes and post Keynesians start from the inductive
axiom that there need not exist a long-run full employment equilibrium in
real world, market-oriented, monetary, production economics' [1980, p. 579],
or Axel Leijonhufvud: 'Many economic decisions are obviously based on induction from incomplete data ...' [1993, p. 6] or Itzhak Gilboa and David
Schmeidler: '[W]hen one engages in inference about the unknown parameter,
one performs only inductive reasoning' [2003, p. 13]. So far, those theorists
who do think equilibrium models should include learning are not providing

---

1. In Boland [1986], I thought this method was just inductive learning but I now
think this also includes what I have been calling induction-like learning.
2. And one that is usually presumed to be an optimal method.

sufficient reason to presume that there is only one type of method – let alone presume (as those examples quoted above seem to presume) that the only type is some version of the inductive method that I discussed in Chapters 6 and 12.[3]

As I discussed in Chapter 9, many macroeconomic model builders in the 1980s and beyond would see building and confirming econometric models as the one method of learning. The idea is that the agents in one's equilibrium model are each conjecturing a model and then learning by using observations to confirm his or her conjectured model. Confirmation of the conjectured model is a matter of degree, which depends on the assessed quality of the model's 'fit' with the observed data. Needless to say, any model with one or more false assumptions can fit a limited amount of available data so it is not always clear what has been learned. As noted in Chapter 12, inductive learning has the same problem. To make either method produce true knowledge, one would have to prove that there does not exist and cannot exist any refuting observational data. It amounts to having to 'prove the negative'. Moreover, accepting an econometrically estimated model as a true explanation when the truth status of its assumptions has not been proven is just an obvious example of exercising the instrumentalism discussed in Chapter 9.[4] Nevertheless, one can still understand why theorists who in the 1970s began to advocate the use of the Rational Expectations Hypothesis were presuming that there is something like an inductive learning method whereby one learns by collecting observational data.[5] As I explained in Chapter 9, such a method would always provide a strong natural connection between learning and collecting information. And the strong connection can be used to explain not only learning successes but also learning failures – any failure to reach a rational-expectations-based equilibrium can easily be explained as the result of an insufficiency in the quantity or quality of available information.

3. With a simple Google-type search, I found the above quotations referring to learning by induction. I also found these: '*inductive* science of economics must start from explicit recognition that every observable action of real life transactors entails finite set-up costs – real or subjective costs that are largely independent of the level of activity to which the observable action is related' [Clower 1994, p. 811] and 'instead of *deducing* optimal actions from universal truths, he will need to use *inductive* reasoning, that is, proceeding from the actual situation he faces' [Vriend 2002, p. 835, emphasis in original]. Question: just what could Davidson [1980] quoted above mean by an 'inductive axiom'?

4. As I explained in Boland [1997, pp. 283–84], econometricians too often unknowingly advocate Friedman's instrumentalism even when they think they disagree with Friedman's methodology.

5. See, for example, Lucas and Prescott [1971], Lucas [1976], Hansen and Sargent [1980], or Spear [1989].

In Chapter 6 I discussed two views of knowledge: the quality-based view and the quantity-based view. The quality-based view is not commonly recognized by economic model builders. In that chapter I presented a discussion of a consumer having to conjecture his or her preference map, and of how, if the conjecture is wrong, the point chosen can possibly not be a utility-maximizing point. Clower's monopolist had a similar problem in Chapter 3. The quantity-based view is commonly presumed in equilibrium models such that in them the amount and quality of information available plays a determining role.[6] The quantity-based view is what is presumed when the model builders think some form of induction is the primary learning method.

There is no inductive proof that there is a necessary connection between the learning process and the prior accumulation of information, and thus relying exclusively on an inductive learning theory is self-contradictory![7] This is of critical importance for the recognition of a role for learning in the process of reaching an equilibrium or in the explanation for the absence of an equilibrium. If knowledge is still considered to be quantity-based, and thus to be obtained either by induction or just by means of econometric model building and thereby mechanically connected to the information collected, it will be virtually impossible to build a theory of equilibrium stability or of disequilibrium which is consistent with the requirements of methodological individualism. This is simply because either type of learning method is considered by many equilibrium model builders to be an objectively 'rational process' that is so reliable that any rational individuals who collect the same information will reach the same conclusions. Unfortunately, equilibrium model builders who presume that one can learn by some type of induction are generally confused about induction and about so-called inductive reasoning. Moreover, those that think econometric estimation is a reliable alternative learning method are just resorting to the questionable instrumentalist methodology that I discussed in Chapter 9.

---

6. Be careful not to confuse the quality of information that the quantity-based theory of knowledge depends on with the quality-based theory of knowledge.

7. It would be difficult to justify the presumption that everyone learns inductively, even on its own terms. How does one learn to learn inductively? If we answer 'inductively', we have an infinite regress; and any other answer admits that learning involves something more than induction. We do know how some people learn to learn with econometric model estimation. They obtain a graduate degree in economics. And, as discussed in Chapter 9, this means model builders who rely on the Rational Expectations Hypothesis are presuming everyone in an economy has the equivalent of a graduate degree in economics! And, more important, in the case of econometrics-based learning, where does the model that is to be estimated come from?

## 15.1. LEARNING VS. KNOWLEDGE
## IN EQUILIBRIUM MODELS

Some readers may think it is my continually associating inductive learning with econometrics-based learning that is confused. So, let me explain why I consider econometrics-based learning to be a version of inductive learning. Supposedly, pure inductive learning first involves making singular observations or collecting singular data with which one would then logically induce some sort of general statement or proposition. The philosopher's classic example involves observing white swans and never a non-white swan and from these observations alone concluding that all swans are white. And on the basis of the assumed truth of the learned proposition that all swans are white, if one were to see in silhouette an incoming bird of the shape of a swan but of a so-far unknown colour, one would predict that the colour of this incoming swan will be found to be white. The problem is that no finite number of observed white swans proves that *all* swans are white. Again, to prove all swans are white, one must prove that there does not now exist a non-white swan anywhere in the world or at any time past or future. Such a proof is impossible even if the statement 'all swans are white' were exactly true.

Unlike pure inductive learning, econometrics does not usually begin with just observations[8] but with a conjectured model for which the values of parameters are initially unknown. One uses the observations to learn the values of the parameters so that when the values of the exogenous variables are inserted into the model, the resulting values of the endogenous variables will 'match their' observed values in accordance with the statistical standards of the day. Again, the question to ask is where does the needed conjectured model come from? As I said, this question is of particular importance for those macroeconomic equilibrium model builders who wish to use the Rational Expectations Hypothesis and then assume all agents in the model use econometrics to form their expectations. This was discussed in Chapter 9, where the idea was that one could assume all agents are econometricians and that all reach the same conclusions whenever facing the same information or observational data.[9]

What makes the econometric method of learning like inductive learning is the source of the equilibrium model that is being estimated. Today few models

8. However, there are some econometrics theorists who claim that one must analyze data before building a model to explain it, to be sure that the assumptions required for performing econometric estimations are not violated by the stochastic nature of the data. See Boland [2014, ch. 10] for an extensive discussion of this aspect of econometric estimation methods.

9. This assumption is explicit in the work of Evans and Honkapohja [2001], discussed in Chapter 9 above. But keep in mind that they are of the newer culture of formal model builders who are willing to assume anything in their model so long as their model yields a stable equilibrium.

being estimated in macroeconomics are new inventions out of whole cloth; rather, they are the result of a long process and are just the latest modification of some previously estimated model. Each step in the process of developing a macroeconomic equilibrium model is in response to resulting confirmation problems revealed by estimating with available data. I am calling this process of developing a macroeconomic model induction-like for the reason that it has the same problems as pure induction. As I noted in Section 9.2 of Chapter 9, to perform the estimations, the model builders presume the parameters of the model are ergodic. The idea that the parameters of any model do not change over time begs too many questions for a realistic equilibrium model particularly when the model is being used to forecast. Similarly, the fact that a model fits the data available today, as some may think can be achieved with induction, does not guarantee it will fit any new data.

Any model one uses to explain available data represents one's current knowledge about what is being explained. With this view in mind, what about the agents in the equilibrium model? What is the agent's knowledge of the situation? Not only must the agent's method of learning be explained in the model, but the agent's current knowledge must also be explained in the model. I suspect that most if not all macroeconomic equilibrium model builders see no reason to distinguish between an agent's learning and knowledge. And failure to distinguish learning from knowledge is the hallmark of pure inductive learning.

## 15.2. LEARNING AND METHODOLOGICAL INDIVIDUALISM

Learning for the purpose of decision making, if it were recognized within a neoclassical economic model, would be a very individualistic activity[10] and thus any commitment to methodological individualism requires a more fully developed theory of learning. Such a theory may require a reconsideration of methodological individualism itself. There is no good reason why any two people facing exactly the same information will reach exactly the same conclusion – even if they were otherwise identical economic agents with identical utility functions or preferences. That is, there is no reason why two individuals would learn the same thing from the same collection of information.

Figure 15.1 illustrates a situation that could be facing two types of learning consumer. Let each be observing a falling price until time $T_0$. The question for each type of learning consumer is whether to buy now before the price starts

---

10. In this context, most economists rule out 'following the crowd' as they would consider this behaviour non-rational. I will relax this perspective below when considering how Keynes viewed decisions based on expectations in his 1937 article.

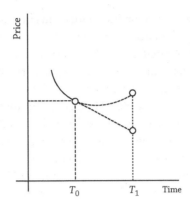

**Figure 15.1.** Expectations of the future price

rising or to wait for it to fall further after time $T_0$. One type of consumer might conjecture a priori that prices cannot fall forever and must eventually rise, such that at time $T_1$ they will be higher. The other might conjecture a priori that the price will continue to fall, such that it will be lower at time $T_1$. The former consumer will buy at time $T_0$ while the latter consumer will want to wait. Yet, as illustrated, up to time $T_0$ the evidence of a falling price is the same for both consumers. The evidence is the same but the conclusions are different simply because the consumers have different conjectured a priori views of price dynamics in general. Without a reliable inductive or induction-like learning method that would preclude the possibility of different conjectured a priori views, such a situation is not unlikely.[11]

I think situations as simple as this lie at the heart of disequilibrium macroeconomics and also seem to be what Hayek was getting at in his 1933 Copenhagen lecture[12] when he explained one possible cause for an economy to be in crisis. Similarly to the neoclassical market dynamics he was discussing, if everyone expects that all prices are going to fall further, there will be a significant deficiency of demand, which yields a self-fulfilling expectation. Likewise, whenever one expects that prices are going to stop falling and start rising, one will find it wise to buy now rather than wait. If everyone expects prices to rise and everyone attempts to act accordingly, prices will be caused to rise by the sudden shift in the demand curve. The issue raised here is not just that expectations matter, but as Hayek explained, that any widespread agreement about expectations can have significant effects on price dynamics. If there really were only one learning method and it was entirely dependent on

---

11. And as I have said, an induction-like learning method such as econometrics is of little use at $T_0$ beyond predicting in accordance with the conjectured a priori views or model.

12. Which I discussed briefly in the Prologue.

the available evidence (and not influenced by any prior conjectures), whenever everyone used the same evidence (such as in our simple example of a falling price) the expectations would be in widespread agreement – and possibly even self-confirming.

If an equilibrium model builder has to depend on such widespread agreement to ensure stable equilibrium prices (as with the Rational Expectations Hypothesis), then recognizing that there is no single reliable learning process may mean that stability cannot be guaranteed – even when the available information is sufficient for what some economists might think facilitates inductive or induction-like learning, as with the illustration of Figure 15.1. Even worse, whenever a model of a persistent disequilibrium is based on a deficient demand caused by a widespread agreement concerning expectations, the absence of a reliable inductive or induction-like learning method means that there is insufficient reason for the persistence of the disequilibrium. The question raised here is why in the absence of a reliable inductive or induction-like learning process would there ever be widespread agreement concerning expectations? Keynes seemed to answer this question by saying we have three different ways of forming our expectations [1937, p. 214]:

(1) We assume that the present is a much more serviceable guide to the future than a candid examination of past experience would show it to have been hitherto. In other words we largely ignore the prospect of future changes about the actual character of which we know nothing.

(2) We assume that the *existing* state of opinion as expressed in prices and the character of existing output is based on a *correct* summing up of future prospects, so that we can accept it as such unless and until something new and relevant comes into the picture.

(3) Knowing that our own individual judgment is worthless, we endeavor to fall back on the judgment of the rest of the world, which is better informed. That is, we endeavor to conform with the behavior of the majority or the average. The psychology of a society of individuals each of whom is endeavoring to copy the others leads to what we may strictly term a *conventional* judgment.

If everyone were basing his or her method or technique of expectation formation on these three alternatives, Keynes claimed that it would lead 'to sudden and violent changes'. That may be true in the long run, but in the short run it may be just the opposite. Let me apply Keynes' alternatives to my simple example – that is, let me briefly drop the common view that all learning must be consistent with methodological individualism. If prices have been falling, Keynes' first conjectured technique leads *everyone* to expect prices to continue to fall. Of course, this cannot go on forever, but how long does it take to get people to stop expecting prices to fall? The second technique does not make sense in the example because falling (disequilibrium) prices may already imply an *incorrect* 'summing up of future prospects'. While the third technique begs the important question about why one individual is less able to form a judgment than the average individual, the short-run outcome is a very stable pattern of behaviour, since everyone is following the same social convention – that is, the same 'rule of thumb'.

The way each individual processes information must be explained if there is no single universally accepted inductive or induction-like learning method, particularly if that information is to be relevant for the formation of expectations.[13] All explanations are matters of the choices being made, particularly in neoclassical equilibrium models. Here there are many questions to consider. How does the individual choose his or her learning technique? If there is a choice to be made, there must be many different techniques (Keynes gave us just three). To what extent does the choice of one technique over another imply a different pattern of behaviour? Differences between techniques must surely imply behavioural differences if learning is to matter at all. Moreover, if there are different learning techniques to choose from and if different techniques imply different patterns of demand or supply decisions, to what extent does the frequency distribution of those techniques over any given macroeconomy's population affect the stability of a neoclassical macroeconomic equilibrium? And finally, if the distribution does matter, how does a model builder explain it without violating the long-standing commitment to methodological individualism?

I think the agenda of any truly improved equilibrium model building in economics must be formed from answers to these questions if such model building is ever going to overcome the inadequacies of the numerous attempts to build models with stable equilibria or models with adequate explanations of persistent disequilibria. And these questions are particularly important for those macroeconomic model builders and DSGE model builders who rely on the representative individual as an acceptable means for providing microfoundations for their macroeconomic equilibrium models.

## 15.3. LEARNING WITHOUT PSYCHOLOGISM OR ANY QUANTITY-BASED THEORY OF KNOWLEDGE AND LEARNING

Let us consider the items on the agenda that I proposed at the end of the last section. The task here will be to reconsider how our usual microeconomic

---

13. Econometric estimation, which is an induction-like method, can vary widely depending on what type of variable is being estimated, and today econometrics theorists continue to argue over the best method in each case.

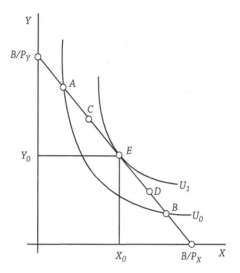

**Figure 15.2.** Choice theory

equilibrium model builder goes about explaining why an observed individual has purchased, say, the quantities $X_0$ and $Y_0$ represented by point $E$ in Figure 15.2, which illustrates the microeconomics textbooks' characterization of consumer choice. Let us start from the usual textbooks' explanation, which says the individual *knows* his or her true utility function or preference map, knows his or her given income or budget $B$ and knows the appropriate prices $P_X$ and $P_Y$. The textbooks would claim the observed individual has chosen the one point on the budget line that maximizes utility or equivalently that provides the most satisfaction because this is where the necessary calculus conditions are fulfilled – namely, where the slope of the budget line equals the slope of the indifference curve through the chosen point. Let us now take a different view. While the prices are public knowledge and the income or budget is in the individual's pocket and thus obviously known, let us again say the individual *does not know* but instead conjectures his or her utility function or preference map.[14] Again the essential question is as always: why did the individual buy point $E$ rather than any other point, such as, say, point $A$? One could answer by claiming that the individual knows that $E$ is better than $A$, but this begs the question of how the individual knows this. Did the individual learn this by trying all possible points such as point $A$? Unless all goods are restricted to discrete quantities,[15] complete knowledge of the utility function is unlikely in

14. As with the discussion in Chapter 6, the main point here will again be that insufficient knowledge of one's preference map is not inconceivable given that the utility or preference map spans what amounts to an infinity of points.

15. That is, a finite number of points – see Boland [1986, ch. 5].

a finite amount of time if learning were really inductive. Without limiting the quantity of any goods to a small finite number, there are just too many points to consider – even along the budget line. In the textbook versions which presume perfect divisibility,[16] complete induction-based knowledge would be impossible because it would require an infinity of trials which would take an unrealistic amount of time.

Let us continue considering the textbooks' paradigm of choice theory, the utility maximizing individual. If the possibility of learning inductively by exhaustive trial and error is effectively denied, what are the remaining options? Following my discussion in Chapter 6, it could be claimed that the individual tried two points, say *A* and *B* of Figure 15.2, and knew that they were not the optimum because in each case the slope of the indifference curve was not equal to the slope of the budget line. This claim, however, would only beg the question of just how the individual knows the slope of the indifference curve. Asking this question does not deny the individual's ability to compare points *A*, *B* and *E* once they have been purchased and consumed. Once purchased and consumed, any point will yield the utility indicated by the true, but otherwise incompletely known, utility function (represented by the preference map shown in Figure 15.2). Thus, after their being consumed, point *E* is then known to be better than either *A* or *B*. But before consuming any of the points, since the individual does not know the true indifference map let alone its slope at any point, how does the individual know point *E* will be the best of all the points between *A* and *B*? How does the individual even know the best point is between *A* and *B*, even when he or she has learned that point *E* is better than *A* or *B*?[17]

As I noted in Chapter 6, what is usually taken for granted is that the individual does not have to learn his or her utility function because it is an exogenous psychological given. This presupposition is much too convenient. Economists concede enough to psychology whenever they claim that the individual can compare two points *a posteriori* on the basis of the derived utilities. Claiming that the individual can compare points that have yet to be consumed goes much too far. This is so even for the individual's perception of the slope of an indifference curve at one point, since in practical terms the slope amounts to

16. Which is needed in the textbooks' version to be able to use calculus derivatives in calculating the slopes of indifference curves.

17. As illustrated in Figure 6.2 of Chapter 6, how would the individual know that his or her preference map is strictly convex to the origin as usually required in textbook explanations of consumer choice? As explained in note 20 of Chapter 6, for the preference map to be strictly convex to the origin, it must be the case that if you draw a straight line between any two points on any of the indifference curves of the map, any point on that line between the two points must be on an indifference curve representing a higher level of satisfaction.

the comparison of two points.[18] Unfortunately, most neoclassical equilibrium model builders believe that a denial of psychology would be a denial of individualism. This belief, which I have called 'psychologism' (see Chapter 6), actually blocks the way to a realistic understanding of individual decision-making. While it may be possible to require that any neoclassical equilibrium model exclude exogenous variables which are non-individualist and non-natural, the identification of the individual with psychological states reduces the role of the thinking individual to that of a simple mechanical link between his or her psychological state (e.g. tastes) and the optimum choice. There is neither autonomous thinking nor free will in a psychology-based conception of the individual's decision making.

As I explained in Boland [2003, ch. 2], a more general version of methodological individualism is needed if psychologism is rejected as unrealistic. The difficulty has always been that economists – starting Schumpeter in his 1909 article – presume that methodological individualism is what the philosopher Joseph Agassi [1960, 1975] calls 'psychologistic individualism'. Agassi also identifies a more realistic version which he calls 'institutional individualism'. Here, for the purpose of my discussion, I want to go further to recognize a more general version of methodological individualism that spans both of these versions. What is important for including what Agassi calls institutional individualism is that it allows for recognizing that individuals are making their decisions constrained not only by natural givens but also by existing non-natural givens that are the result of *past* decisions both by the individual whose behaviour one might be trying to explain but also by the past decisions of other individuals.[19] These constraints are merely the results of the history of past decisions, much like how the evolutionary and Santa Fe economists would see the situation facing an individual's decision making. And as such, the constraining institutions are not beyond explanation and hence not truly exogenous.

As I argued in the Prologue, the primary reason for building equilibrium models is that the concept of equilibrium allows for individuals to be making decisions freely (i.e., autonomously) yet still permits us to explain the state of an entire economy. If my argument is correct, psychologism needs to be avoided in neoclassical models of stable equilibria or of disequilibrium process analysis. Instead the individual needs to be seen to be either, unrealistically,

18. To say otherwise brings up some age-old difficult questions concerning the realism of infinitesimals and similar problems about the realism of calculus – see Boland [1986, ch. 5].

19. The most obvious would include *existing* institutions such as the market structure or more important, past consumption choices of the individual which reduce the size of the remaining budget.

learning inductively his or her indifference map from experience alone, or, more realistically, forming a conjecture about the map. Inductive learning, without the help of some sort of conjectures (econometric, Bayesian, or otherwise) faces insurmountable problems in real time. If realistic models of stable equilibrium or of disequilibrium process analysis are going to be built, then model builders must drop the reliance on inductive learning in the explanation of all expectation errors. And if they instead choose to assume the agents in their models rely on econometric estimation, then they need to explain the source of the agents' conjectured econometric models to be estimated. Either way, some form of autonomous *conjectural knowledge* needs to be considered.[20] Unlike induction-based knowledge, where presumably any insufficiency is due to problems with quality or quantity of information, conjectural knowledge has the potential of being wrong in many more ways. To deal with learning and the formation of expectations using conjectural knowledge, model builders must come to grips with the many difficult questions of methodology.

Returning again to the simple textbook paradigm of utility maximization illustrated in Figure 15.2, let us continue to take prices and income as known givens. For those who want to improve microeconomic equilibrium model building, I think it would be wise to consider the individual decision-makers who do not know a priori their indifference maps, even though there are true maps to be learned if they had an infinity of time. That is, if consumers could try every point on the map, their true maps could be plotted (by connecting the points with the same level of utility obtained). But since generally, as I have often noted in this book, that would be an impossible task in real time and thus each consumers must form a conjecture about his or her map. Each trip to the market is, then, a test of the individual's conjecture, much as it was in Clower's 1959 article that discussed in Chapter 3. His Model III viewed each of the ignorant monopolist's trips to the market as being a test of its assumption concerning the market's demand curve. Such a test of one's assumptions concerning elements of decision making immediately raises a primary question which is of methodological importance. What will be the individual consumer's response to a test which refutes his or her conjectured indifference map? It is most important to note that when individuals have to base their demand or supply decisions on conjectures or assumptions, whatever would have been considered to be a market equilibrium or disequilibrium is put into a different light. Even if the market clears today, unless the individuals are satisfied that their respective maps have been correctly conjectured, market clearance does not directly imply an equilibrium or an optimization (as we saw with Clower's ignorant monopolist). If any individuals think they have

20. And as will be discussed later, since it is conjectural, the individual may occasionally choose to test it.

made a mistake even though the market cleared, there is no guarantee that the market will clear the next day.

## 15.4. EQUILIBRIUM STABILITY AND ACTIVE LEARNING

The assumption that decision-makers form conjectures does not for the most part dramatically affect our concept of a market equilibrium. Consider again an individual consumer who does not know his or her indifference map and thus has formed a conjecture about the map. Let us say this consumer has made the trip to the market and is successful in purchasing the quantities planned. If the individual only has a conjecture about his or her indifference map and there have been relatively few trips to the market, how does the individual know that the chosen point is the one which truly maximizes utility? In simple terms, all that the individual has learned is the level of utility achieved for the chosen point, but he or she does not know whether that level is the best possible for the given budget since full knowledge of the map would at least require a very large number of trials.

There are obviously many ways to try to determine whether utility has been maximized – each depending on the individual's method of learning through trial and error.[21] How much or what kind of evidence would it take to convince the individual that his or her conjecture is correct? Going further, how does an individual learn whether his or her tastes have changed? This second question puts in doubt even states of equilibrium where all individuals are convinced that their conjectures are correct that day. Given that textbooks represent tastes only with indifference maps, if it is admitted that tastes might change, how does the textbooks' individual know they have changed without trying points which are not optimal according to the currently conjectured map? Both questions raise an important issue. The neoclassical equilibrium model presumes that every individual is choosing the point which maximizes utility, but here I am suggesting that from time to time the individual might deliberately choose a *conjectured sub-optimal* point to test either whether the currently conjectured map is true or whether tastes (and thus the map) have changed. If such perverse behaviour is possible, what are the implications for neoclassical equilibrium models?

It might at first seem that such perverse behaviour would be devastating for the usual neoclassical maximization hypothesis, but it has some constructive implications for equilibrium stability analysis. To see this consider again the conceivable market configurations illustrated in Figure 15.3 (which is a repeat of Figure 7.2 to make it easier to discuss here). As I said in Chapter 7,

---

21. See further Boland [1992, ch. 11].

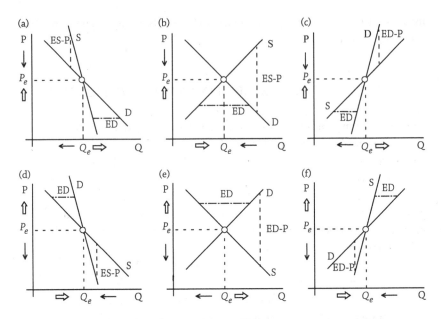

**Figure 15.3.** Possible markets (same as Figure 7.2)

individuals facing a disequilibrium can adjust the price in response to an insufficient demand or supply (i.e. Walrasian price-adjustment behaviour[22]) and can also adjust the quantity in response to a difference between the demand price and the supply price (i.e. Marshallian quantity-adjustment behaviour[23]), the only configuration that has the possibility of ensuring equilibrium stability is the market presented in most textbooks. Namely, only when the market is characterized by downward sloping demand curves and upward sloping supply curves, as illustrated in Figure 15.3(b), is it possible to conceive of individuals facing a disequilibrium and making autonomous and independent decisions that constitute stabilizing responses (i.e. convergence toward market clearance). If in Figure 15.3(b) the price were, for any reason, *not* the market-clearing price, the responses of the individuals facing the disequilibrium will (so long as they are small adjustments) always be in the right direction. Small adjustments will never be destabilizing, that is, never cause a greater discrepancy between demand and supply. This is obviously not so in the worst possible case, Figure 15.3(e), where both curves are sloping in

22. This behaviour was explained by the 'economics teacher' in the dialogue presented in Chapter 7.
23. Similarly, as explained in Chapter 7, this is the textbooks' theory of what the profit-maximizing firm does when the market price offered by demanders differs from the firm's profit maximizing supply price – that is, from its marginal cost – for the current supply quantity.

the wrong direction; both ways of responding to a disequilibrium in this case would make the disequilibrium worse.

For anybody who wishes to argue that the competitive market-system for social co-ordination should be relied on, things can still be problematic.[24] Whenever people can opt for either Walrasian or Marshallian adjustment responses, unless the market appears like Figure 15.3(b) it would be difficult to argue that prices are informative and thereby always promote market equilibria. Whenever the market is otherwise than 15.3(b), there is always the possibility that either type of response behaviour could be destabilizing. In the worst case, Figure 15.3(e), unless demand just happens accidentally to equal supply when the market opens, either type of response behaviour will cause the price to rise or fall at an increasing rate – that is, the market would virtually explode or collapse. A similar problem arises when the demand and supply curves slope in the same general direction. Again, if the market is configured like Figure 15.3(a) or (c),[25] Walrasian adjustment behaviour is stabilizing but Marshallian adjustment behaviour is destabilizing. Thus, whether the market is stable depends on which behaviour dominates in the market. I will leave this macroeconomic question for now. Unless one has reasons to ensure that all markets are like Figure 15.3(b), one needs to be very careful about recommending complete dependence on the market system as a means of organizing or coordinating society.

If the price were accidentally set at the market-clearing price in the usual neoclassical equilibrium model, (perhaps with the help of an auctioneer), there would be no need to worry about the relative slopes of the demand and supply curves. When facing such a market clearing price, all demanders and suppliers would make the required maximizing decisions and thus would have no reason to change their demand or supply quantities. Moreover, in such a case, the question of stability would not arise – even if the market were like Figure 15.3(e)! But this is only because in the usual neoclassical equilibrium model all individuals are assumed to be making the correct decisions and to know that they are making the correct decisions. As I have noted, any shift from equilibrium would, however, cause an explosive disequilibrium if the market happened to be as shown in Figure 15.3(e).

24. See the dialogue in Chapter 7, as well as Hayek [1945].
25. While some might claim that these configurations are precluded by ordinary market behaviour, they should be reminded that it is not uncommon for sellers to offer quantity discounts and thus have downward sloped supply curves. There is also the so-called Veblen effect that refers to purchases of ostentatious luxury goods for which when the price goes up, so does demand. And there is the Giffen effect that refers to low priced goods for which the demand goes up when the price does because it diminishes the consumer's purchasing power. The demand curve for these 'Giffen goods' can thus be positively sloped. For more, see Boland [1992, ch. 14].

Let us now consider Figure 15.3(e) in the context of the modified neoclassical equilibrium model that I was discussing in Section 15.3 – that is, the model in which all decision-makers base their decisions on conjectured objective functions.[26] While the market may clear one day, there is no guarantee that it will clear the next day, even when the observable evidence is the same on both days. As suggested in my simple example of the utility maximizer, a consumer may try a sub-optimal point the next day to test his or her conjectured map. If the market were like Figure 15.3(e), such a test would cause the demand curve to shift from the one which intersected the supply curve the day before.

If psychologism is rejected and people in the equilibrium model are allowed to test their conjectures *and* allowed to respond to disequilibria with either Walrasian price-adjustment or Marshallian quantity-adjustment behaviour, the only type of market that will *guarantee* equilibrium stability is, as I have been saying, the textbooks' market with a downward sloping demand curve and an upward sloping supply curve, i.e. Figure 15.3(b). Model builders must make such allowances if individuals are allowed to be free to make any decision they wish or to test their decisions. This means that if the model purports to explain a market in the world one can see out their window and it is a world truly populated by autonomous individuals as all neoclassical model builders seem to think, then that market must be one like Figure 15.3(b) since if any other market configuration existed, *that market would have exploded or collapsed by now*.

This argument for the textbook equilibrium stability case is rather indirect and depends heavily on the predominance of the maximizing behaviour usually assumed in neoclassical equilibrium models. That is, only a small proportion of the market participants can be engaged in perverse testing of their conjectures, otherwise the usual concept of demand and supply curves would lose their meaning. If there is a small proportion of the market participants who are actively testing their conjectures everyday then the only market that can be in a stable equilibrium is the textbooks' market. These considerations emphasize the need to consider the macroeconomic question about which response behaviour dominates the whole economy. In configurations other than Figure 15.3(b), why should the stabilizing response dominate rather than the destabilizing response whenever people can behave in either way?[27]

26. For those readers unfamiliar with the term 'objective functions', this is mathematical jargon which refers to the mathematical representations of such things as utility functions or functions resulting from the calculation of profit based on a production function.

27. Interestingly, Hicks [1956] recognized this dilemma when considering the possibility of markets with upward sloping demand curves. He was specifically addressing the possibility of a Giffen effect that I noted in note 25 but he claimed that it would be unlikely and thus dismissed its relevance. But, as I am arguing here, this only begs a macroeconomics question about the distribution of these 'unlikely' consumers.

## 15.5. ARE MACROFOUNDATIONS NEEDED
## FOR EQUILIBRIUM MICROECONOMICS?

Providing microfoundations for macroeconomics equilibrium models has been a concern for more than four decades – particularly in terms of the adequacy of explanations of disequilibrium phenomena.[28] In many ways this may be missing the point. It can be argued that one thing that can be learned from Keynes is that neoclassical model builders lack macrofoundations for their microeconomic equilibrium models.[29] But surely the question of the equilibrium stability of the whole market also raises the macroeconomic questions noted at the end of the previous section. Which response behaviour dominates the whole economy must be explained using some sort of perspective on the economy as a whole which cannot be deduced from the behaviour of individuals alone and certainly not using a representative agent. Even if model builders were to allow for significant diversity in terms of different types of rational response due to differing individual aims, they still must explain the macroeconomic distribution of those aims to ensure equilibrium stability.

I considered a related issue in Boland [1986, ch. 6]. Hayek in his 1933 Copenhagen lecture pointed out that explaining a crisis or just a state of disequilibrium might require claiming widespread expectational errors, thus begging the question of why so many people could be wrong.[30] And as I noted above, in his 1937 article Keynes raised questions of how individuals facing 'uncertainty' form expectations, and he provided three conceivably different ways to form expectations. While Keynes argued that all of his noted ways were destabilizing in the long run, I argued earlier in this chapter that they may be stabilizing in the short run. Also, it could be argued that his potentially destabilizing techniques are destabilizing only if they necessarily lead to false expectations. There is no reason why someone using one of Keynes' techniques of expectation formation cannot happen occasionally to form correct expectations. To think uncertainty necessarily leads to incorrect expectations is merely an expression of a belief in inductive learning. Nevertheless, whether the use of any particular technique of forming expectations is stabilizing or destabilizing may depend on how widespread is its use. Unfortunately Keynes' only explanation of how widespread is the use of his three techniques was entirely based on his 'psychological laws'.

I did explain how expectations can be self-fulfilling in the simple case of price dynamics depicted in Figure 15.1 earlier in this chapter. Clearly it matters whether everyone expects prices to continue falling, but what happens

---

28. See Hoover [2012].
29. See Boland [1982, ch. 5].
30. In his 1933 lecture Hayek was saying it was caused by government policies that interfered with the market by changing the interest rate.

when expectations are mixed? This question is not addressed in the equilibrium stability literature because of the tendency to think that there is only one technique of expectation formation. If model builders were to address this question in terms of equilibrium stability analysis,[31] they would have to ask why rational decision-makers would have different techniques. As I have been suggesting, neoclassical equilibrium model builders will often claim there is only one technique and it is psychologically given. Obviously, I am continually pushing this line of questioning because I think that not only is there more than one conceivable technique, but also that the techniques are not psychologically given. It is time now to present the problem of expectation formation in different terms.

## 15.6.  CONJECTURAL KNOWLEDGE AND ENDOGENOUS EXPECTATIONS

Throughout I have be arguing that there is no reliable inductive learning method or inductive-like method such as econometric estimation, and that without conveniently presuming psychologically given tastes and techniques, model builders should assume individuals make decisions on the basis of theories they conjecture to be true. There are many theories involved in any decision. The simplest theories are those about price expectations. Consider again the example of Figure 15.1 and let us ask various individuals why they think the price will rise or continue to fall after time $T_0$. One individual may believe in the theory of inductive learning and say that the reason prices are expected to fall is that they have been observed to be doing nothing but falling for some time. Another individual may believe in an a priori theory that average prices are determined by real costs and that the daily price may oscillate about the average such that whenever the price falls for a while it will surely rise to restore the average. The question of widespread agreement over expectations then becomes a question of widespread agreement over either inductive learning or a priori conjecturing about price oscillations. Perhaps a strong argument could be made that this is a question about the sociology of knowledge. It is certainly not a question about any differences between the 'information sets',[32] as many of the equilibrium stability theorists might think, since in my Figure 15.1 example the information set is the same for both individuals prior to time $T_0$.

31.  See Pindyck and Rubinfeld [1998, ch. 14].
32.  Again, by 'information sets' I am simply referring to collections of observational data or observation reports.

If it is allowed that expectations can be based on conjectural knowledge, this idea can easily be extended to all decision-making processes involved in neoclassical economics.[33] The individual consumer can operate on the basis of a conjectured theory of learning, a conjectured indifference map, a conjectured theory of how to change the givens such as prices or incomes, and so on. Whenever the question of providing reasons for equilibrium attainment actually requires dealing with learning or induction-like learning such as econometric estimation, the basis for equilibrium explanations will thus have to be extended to explain how individuals learn or conjecture these theories or econometric models. By insisting that there is no universal learning method such as induction, and that the needed theories are not psychologically given, I am insisting that there is no reason why everyone should agree about any one operative theory, let alone all of them. To the extent that widespread agreement matters for either equilibrium stability or persistent disequilibria, there is certainly a need to explain the extent of the agreement. Such an explanation would be an important beginning for providing macrofoundations for microeconomics.

## 15.7. GENERALIZED METHODOLOGICAL INDIVIDUALISM

From the beginning of this book I have argued that the dominance of equilibrium model building in neoclassical economics can be understood by seeing how the idea of an equilibrium allows social coordination while simultaneously allowing free-willed or autonomous independent decision-makers. As such, a neoclassical equilibrium model is designed to foster methodological individualist explanations of the economy. In such explanations only individuals make decisions and they make them while being ultimately constrained only by the limits of Nature and guided by only their own personal aims (without saying where these aims come from). Two individuals facing the same information and circumstances may still make different decisions if they have either different aims or different conjectures about their circumstances. Moreover, if it is allowed that individuals make conjectures in their decision making process, then a generous source for diversity will have been provided beyond differing constraints or aims.

While over the last seventy years most neoclassical equilibrium model builders have been satisfied to think that learning and maximizing are processes beyond question, some have noticed that inductive and induction-like

---

33. Recall again that Clower's ignorant monopolist engaged in conjecturing about the slope and position of the demand curve it faces.

learning methods are not reliable whenever the available information is quantitatively or qualitatively inadequate (see Chapter 9). Some critics have instead been quick to abandon equilibrium methods on the grounds that learning or maximization would require too much of any available information (see Chapter 11). Other model builders have instead seen all this as an interesting puzzle to be solved. How can one assume the economy is in equilibrium when there are insufficient grounds to assume that the individual participants are capable of making perfectly optimizing decisions that are perfectly consistent with a state of equilibrium (see Chapter 12)?

Obviously there are some optimistic model builders that will still like to create a workable version of methodological individualism that is consistent with both psychologism and inductive or induction-like learning theory. Can model builders reject the notion that explaining all or any individual's behaviour must be reducible to a psychological explanation and still maintain that all learning is ultimately inductive or inductive-like? If it is assumed that rational decision making does not ultimately have to be explained only in terms of psychology (thereby rejecting the psychologistic version of methodological individualism), perhaps it also could just be assumed that rational decision making itself is learned. If equilibrium model builders assume that individual decision makers learn what is needed to make rational decisions inductively, it still could again be asked whether they learn the presumed inductive method of learning itself inductively. That is, do people learn to be inductive?[34] As I suggested at the top of this chapter, the infinite regression here should be obvious.[35]

Maybe model builders should instead admit that psychologically given learning methods are informationally insufficient for providing accurate expectations for correct 'rational' decisions. If they admit this, individuals could be assumed to know how to deal with such an insufficiency of the information basis for their expectations. The old Rational Expectations Hypothesis was invented to close this circle. Expectations were to be rationally chosen like anything else – perhaps, it was said, the rational decision process is more like a combination of econometrics and ordinary calculus. The circle would thus be closed without having to give up on the assumption that adequate psychological skills are exogenously given.

34. In fairness it should be recognized that to a certain extent people do when their grade school teachers try to convince them of some idea or proposition by giving them numerous examples.

35. As I suggested in note 7 above, for econometrics-based learning, economics students do learn how to use econometrics to learn, of course, but the infinite regress does not apply unless teachers use examples of econometric learning to demonstrate how to learn econometrically.

There is still a problem that some less optimistic model builders may wish to address. If equilibrium models are to be the basis of explanation in economics, the possibility of individual behaviour being consistent with a state of equilibrium does not ensure such an equilibrium can exist. The idea of an equilibrium is not captured by static properties of a single point, such as the one where demand equals supply, since such an equation only amounts to a balance. As was discussed in Chapter 2, it must also involve a reliable dynamic process of reaching or maintaining that balance point. In this case, providing such a process in the model means that it is a stable balance. An equilibrium is just that, a stable balance and thus the textbooks' idea of an unstable equilibrium is self-contradictory! Going further, not only does the neoclassical idea of an equilibrium require an explanation of the dynamic process of reaching the equilibrium, what is most important and what I have been stressing is that the explanation cannot violate the requirements of methodological individualism. After all, allowing autonomous and independent individuals to all be simultaneously optimizing is why the idea of an equilibrium has long been interesting in the first place (see the Prologue and Chapter 6).

How individuals learn independently to make decisions in a manner that is unintentionally stabilizing has been the main puzzle facing the explanation of the process of reaching the equilibrium. It is unfortunate that the solution of this puzzle is still thought by many to require a fuller understanding of the psychology of learning (as with the Santa Fe researchers I discussed in Chapter 12). The stability of an equilibrium model based on expectations which individuals are said to have learned inductively is either a false stability or a false individualism, since the stability of the equilibrium attainment is due only to a presumed and false theory of learning – which, as in the case of the Rational Expectations Hypothesis, is that any two individuals facing the same information are thought to form the same expectations.[36] So, the problem of equilibrium stability analysis must be seen as a problem, not only of satisfying methodological individualism by having only individuals make decisions, but also of allowing the individuals to be autonomous. But in the absence of a reliable inductive or a perfect econometric learning process, why would any two autonomous individuals always be forming the same expectations as claimed by the Rational Expectations Hypothesis? The problem of equilibrium stability needs to be seen as that of how a market's balance can be stable when individuals facing the same information are systematically forming *different* expectations.

36. As I have been saying, those macroeconomic model builders that use the Rational Expectations Hypothesis usually presume that everyone is an econometrics learner but still presume the agents will all form the same expectations allowing for stochastic variation only.

One might ask, does the possibility of autonomous and diverse individualism necessarily lead to chaos or anarchy? The reason for the stability of equilibrium attainment in a present day economy may just be that there is considerable homogeneity in the accepted views of learning and proper behaviour – even though the homogeneity is not a psychological given. The homogeneity may simply be a matter of sociology – that is, the individuals' reliance on existing social institutions to make decisions and choices.[37] Many neoclassical theorists might be interested in building equilibrium models which allow both diversity of individual aims and the possibility of a methodological-individualist explanation of prices. Doing this is all too easy since diversity is always provided by the common assumption that everyone is to some degree different – that is, no two people are exactly alike.[38] In this case, such diversity is not explained – it would just be assumed to be an exogenous variable in such a neoclassical equilibrium model. However, inventing a new technique of equilibrium model building with exogenous diversity does not constitute the creation of a different form of microeconomics – instead, it amounts to a microeconomics with some needed macrofoundations which in this case could be based on existing social institutions and related matters of sociology that might explain the nature of the diversity.

Any equilibrium model builder preoccupied with building models with exogenously diverse tastes could easily overlook a far more important intellectual challenge involving both social institutions and sociology in general. The major theoretical task for neoclassical economics equilibrium model building is not to explain why people make different choices when they are given the same information but why so many of them make the same choices when there is so much room to be different. Such homogeneity is endogenous rather than psychologically exogenous. The stability of any given society or economy may be due to individuals choosing to conform rather than being due to the independent, autonomous individualism on which most economists like to base their neoclassical equilibrium models. If, as I would hope, model builders are looking to overcome all of the old, long-standing problems with microeconomic equilibrium models that I have been identifying and criticizing in this book, I think a truly new and improved microeconomics must not then make the autonomy of individualism psychologically exogenous but make it also a matter of sociology-influenced choice at least to some degree. This surely involves the possibility of heterogeneous learning techniques. While

---

37. See Boland [1992, ch. 8] where I have provided a theory of institutions which asserts that social institutions are embodiments of solutions to social problems and exist to enable individuals to make decisions and choices both social and personal.

38. Of course, monozygotic twins start out very much alike but things usually change as they grow up.

allowing for disequilibrium responses and allowing that decision plan formats and diverse tastes are important, model builders should not be satisfied with models that fail to provide a truly endogenous basis for the questions of either equilibrium stability or persistent disequlibria. And of course, the problem of equilibrium stability must rightfully occupy center stage for anyone interested in intellectually consistent neoclassical equilibrium models within which autonomous individualism really matters.

# Epilogue

## *Prospects for changing equilibrium model building practice in economics*

In this Epilogue I will be considering whether new efforts provided by the behavioural and the evolutionary as well as complexity economics that I discussed in Chapters 11 and 12 have a chance of changing how economics and economic model building is practiced in the future. And I will also consider whether any such change can affect how economics is taught to beginning economics students.

### E.1. BEHAVIOURAL AND EXPERIMENTAL ECONOMICS AND EQUILIBRIUM MODELS

Thanks to Herbert Simon's 1957 book and his famous 1955 article on behavioural economics, there is the standard presumption that being behavioural must involve psychology. And the researchers associated with the Santa Fe Institute have contributed to the perpetuation of this presumption. But it is still a presumption. Many of the questions raised at Santa Fe which were discussed in Chapter 12 (viz., those concerning such things as learning, increasing returns and path dependency) do not have to be about psychology. It can just as easily be a matter of sociology. Either way, among some mainstream model builders there has been considerable reluctance to consider psychological considerations,[1] as is evident in Stigler and Gary Becker [1977].

---

1. Tibor Scitovsky [1976] was an exception in his day.

For Simon in the 1950s, the main characteristic of behavioural economics was the notion that psychology should somehow matter when explaining
human decision making. Simon was challenging how we teach microeconomics, of course. From his perspective, neoclassical equilibrium models simply
lack a psychological perspective.[2] He particularly questioned the presumed
inherent human cognitive abilities of the neoclassical consumer or producer.
Although he did not talk about equilibrium models there, in effect he was
claiming that it is impossible for any individual to actually do everything psychologically necessary to assure the achievement of simple utility maximization that is central to neoclassical equilibrium models. As Simon put it [1955,
p. 101]

> Because of the psychological limits of the organism . . . actual human rationality-
> striving can at best be an extremely crude and simplified approximation to the
> kind of global rationality that is implied, for example, by game-theoretical
> models. While the approximations that organisms employ may not be the best –
> even at the levels of computational complexity they are able to handle – it is
> probable that a great deal can be learned about possible mechanisms from an
> examination of the schemes of approximation that are actually employed by
> human and other organisms.

Today, behavioural economics is mostly of interest to those economists
engaged in conducting laboratory experiments that test the behavioural assumptions of many mainstream equilibrium models. Various regularities
of behaviour found in their laboratory experiments are claimed to deny the
truth status of what is assumed in common economic equilibrium models.
An early and well-known example of such denials involved so-called 'preference reversals' [Lichtenstein and Slovic 1971]. One very important example,
Daniel Kahneman and Amos Tversky [1979], directly tested the now common
assumption used in DSGE models when the model's agents' choice behaviour
is characterized by assuming they maximize expected utility. The test specifically found that consumer's choices can vary depending on how the choice
questions are framed.

## E.2. BEHAVIOURAL AND EVOLUTIONARY ECONOMICS AS ALTERNATIVES TO EQUILIBRIUM MODELS

Reading the recent literature by evolutionary economists one gets the impression that they are more concerned with the survival of their proposed

---

2. He was not concerned with the psychological basis of preferences but instead
the psychological basis of decision making.

paradigm than with whether it can replace or even just repair mainstream neoclassical equilibrium economics. As Winter [2014, p. 629] complains:

> Evolutionary economics is oriented to the system level (or the 'population' level) from the start, and is not encumbered by commitments to fiction at lower levels of analysis (organism, individual, organization). Its actual commitments encourage precisely the engagement with reality that many observers find lacking in the mainstream. Why then, does the evolutionary approach struggle to gain attention while the visible 'talking heads' offer policy conclusions based on theoretical arguments that nobody actually believes? . . .

He goes on further to suggest [2014, p. 638],

> Evolutionary economics has certainly not achieved critical mass there, and is much weaker even in its areas of relative strength than it would be if the movement were broader. The probability of survival, let alone success, might actually be enhanced by a more broadly expansionary policy.

And Ulrich Witt seems to agree [2014, p. 660],

> The heterogeneity of theories and topics associated with 'evolutionary economics' creates the impression of a fragmented research field. In its present state, evolutionary economics seems far from living up to the initial expectations of some of its proponents, namely that an evolutionary approach could mean a paradigmatic shift in economics.

Some behavioural economists do not always see their job as that of replacing equilibrium neoclassical economics models. Instead, many are trying to help by filling out the meaning of the behavioural assumption of maximization at the center of mainstream economics. Others, particularly those engaged in so-called behavioural finance actually see behavioural to mean non-equilibrium.

## E.3. COMPLEXITY ECONOMICS AND EQUILIBRIUM MODELS

It is not always clear what a Santa Fe researcher such as Arthur wants to happen with mainstream equilibrium model building. As he [2014, p. 2] observes

> Over the last twenty-five years, a different approach to economics has been slowly birthing, and slowly growing – *complexity economics*. . . . What does this different way of thinking about the economy offer? How exactly does it work

and where does it fit in? Will it replace neoclassical economics, or be subsumed into neoclassical economics?

On the one hand, as discussed in Chapter 12, they identify major problems with such model building. Most of the problems concern missing elements necessary for realistic models of the economy. As Arthur [2014, p. xix] explicitly claims in the Preface to his book on complexity economics, from the perspective of his complexity economics a realistic model of the economy would not necessarily be an equilibrium model. Instead, any realistic model would recognize that the economy is usually 'in nonequilibrium'. In such a model, the decision makers are constantly changing their actions in response to the outcomes of their decisions. Moreover, the outcomes will usually not be the result of their actions alone but be mutually created by all participants in the economy.

Arthur concedes that general equilibrium theory is, of course, mathematically elegant. And he allows that a general equilibrium model lets us have a picture of the economy and 'to comprehend the economy in its wholeness' but this comes at a price that economists should object to if they want a realistic model. Given what we saw with the Great Recession of 2007 and 2008, relying on equilibrium models makes it difficult to anticipate such system breakdowns. Presumably, he thinks his complexity economics could anticipate break downs but non-complexity economists will likely have difficulty seeing how that is possible.

In his introductory chapter, Arthur [2014, p. 25] concludes:

> Complexity economics is not a special case of neoclassical economics. On the contrary, equilibrium economics is a special case of nonequilibrium and hence complexity economics. . . . Equilibrium of course will remain a useful first-order approximation, useful for situations in economics that are well-defined, rationalizable, and reasonably static, but it can no longer claim to be the center of economics.

## E.4. TEACHING WITH EVOLUTIONARY OR COMPLEXITY ECONOMICS

Whether one considers the aggressive program of the Santa Fe economics researchers or the longer-standing patient plea of the evolutionary economists for an appreciation of their program, it is unlikely that teachers of the ubiquitous Economics 101 will jump on either bandwagon. A simple reason is that one cannot appreciate what either offers since both are limited to responding to the state of neoclassical equilibrium model building. This means that, until students are well versed in neoclassical model building, they are unable to appreciate that there are problems that need to be dealt with. And too often students will object to learning about problems of something they are supposed to learn on its own.

From another perspective, we see that such problems are sometimes discussed in graduate programs, since after all, those students as undergraduates have had to learn the neoclassical equilibrium model. But, even in the graduate program, it is more likely that a problem is only discussed after the teacher thinks a solution exists. Despite such reluctance at the graduate level, one can find elective undergraduate classes at the third and fourth year levels that involve behavioural and experimental behavioural economics, so maybe there is hope for the evolutionary and experimental programs.

## E.5. THE ISSUE OF LEARNING MUST BE DEALT WITH

From the beginning of this book I have been saying that the main and most important missing element of neoclassical economics equilibrium models is recognition of how the agents – whose behaviour the models is explaining – learn what they need to learn for an equilibrium to be reached. Needless to say, I am not the first to be saying this. After all, that was Hayek's main point in his 1937 article, 'Economics and knowledge' and Richardson's main point in his 1959 article discussed in Chapter 4.

While it is easy for me to say that equilibrium models need to address the question of how the models' agents learn what is needed for equilibrium, it is quite different when it comes to actually answering that question. Arthur recognizes a key difficulty. Whether one is dealing with learning directly or indirectly by recognizing increasing returns, without an identifiable optimum way of learning, the result of realistic learning methods will usually be path dependent. Path dependency means two things. On the one hand, unlike what one can do with a formal mathematical equilibrium model – with which one can predict the equilibrium conditions and values for all of the endogenous variables such as the value of the equilibrium prices (or more likely, the equilibrium relative prices between different commodities) to engage in comparative statics analysis – path dependency means that the best we can do is discuss historically the evolution of reaching the path dependent outcome. On the other hand, there is no single method for explaining the path taken. In Arthur's view, the path may depend on unpredictable random events so the outcome must be addressed case-by-case, which makes the path dependency approach hardly something to compete with general equilibrium model building – particularly whenever one wishes to answer questions of the form: 'What exactly will happen to endogenous variable $Y$ when exogenous variable $X$ changes by, say, 10%?' These are the types of question a government policy maker tends to ask. But Arthur is not alone in identifying the path-dependency result. For example, Fisher [2003, pp. 79–80] notes

> When trade takes place out-of-equilibrium (and even more when disequilibrium production and consumption occur), the very adjustment process alters the equilibrium set.... This is easily seen even within the simplest model of pure exchange. In such a model, the equilibrium prices and allocations depend on the endowments. If trade takes place out-of-equilibrium, those endowments change. Hence, even if the trading process is globally stable, the equilibrium reached will generally not be one of those corresponding to the initial endowments in the static sense of the Walras correspondence. Rather the equilibrium reached will be path-dependent, dependent on the dynamics of the process taking place in disequilibrium.

Evidently Fisher similarly thinks that out-of-equilibrium behaviour – such as when the market participants are trying to learn or find the equilibrium price by trial and error – can easily lead to an equilibrium state which is path-dependent even in a stable Walrasian general equilibrium model.

There is also the matter of learning itself. As I explained in Chapter 12 and noted often in Parts II and III, despite what many economic model builders believe, there is no inductive learning method and we have known this since the end of the eighteenth century. While econometric learning is obviously possible, the results cannot compete with a reliable inductive learning. Of course, if there were a reliable inductive learning method, we could assume everyone has available sufficient information or data that would allow them to assuredly reach their personal equilibria. And, even if there were a reliable learning method, having sufficient information available is not only unlikely but impossible – particularly when it comes to data about the future. Of course, again, I am not the first to say this as that was one of the main points of Keynes 1937 article, 'The general theory of employment', that I discussed in Chapters 14 and 15. The main point with all of this is simply that without something like a reliable inductive or perfect econometric learning method, decision makers are instead relying on conjectures about what any data means; but there is no method, obviously, for optimally conjecturing – however convenient such a method would be for equilibrium model builders.

It is hard to deny all of the problems with equilibrium models that I have been discussing in this book. Some of those problems are recognized even by equilibrium model builders. Some model builders might accept the Santa Fe researchers list of problems, too. But if the problems are ever going to be solved, it will not be done by outsiders to the mainstream. It will take the equilibrium model builders themselves to finally address the problems – including the problems involving realistic learning in their models. Hopefully, there are some equilibrium model builders out there that will agree with what I have been saying in this book and thus think the equilibrium model building method can be improved by addressing the problems discussed in this book.

# BIBLIOGRAPHY

Agassi, J. [1960] Methodological individualism, *British Journal of Sociology*, 11, 244–70.

Agassi, J. [1975] Institutional individualism, *British Journal of Sociology*, 26, 144–55.

Albert, M. [2001] Bayesian learning and expectations formation, in Corfield, D. and Williamson, J. (eds.), *Foundations of Bayesianism* (Boston: Kluwer), 341–62.

Alchian, A. [1950] Uncertainty, evolution and economic theory, *Journal of Political Economy*, 58, 211–21.

Anwar, S. and Loughran, T. [2011] Testing a Bayesian learning theory of deterrence among serious juvenile offenders, *Criminology*, 49, 667–98.

Arrow, K. [1959] Towards a theory of price adjustment, in Abramovitz, M. (ed.), *Allocation of Economic Resources: Essays in Honor of Bernard Francis Haley* (Stanford: Stanford University Press), 41–51.

Arrow, K.J. [1974] General economic equilibrium: Purpose, analytic techniques, collective choice. *American Economic Review*, 64, 253–72.

Arrow, K. [1994] Methodological individualism and social knowledge, *American Economic Review: Papers and Proceedings*, 84, 1–9.

Arrow, K.J. [2005] Personal reflections on applied general equilibrium models, in Kehoe, T.J., Srinivasan, T.N. and Whalley, J. (eds.), *Frontiers in Applied General Equilibrium Modeling, In Honour of Herbert Scarf* (Cambridge, UK: Cambridge University Press), 13–23.

Arrow, K. and Debreu, G. [1954] Existence of an equilibrium for a competitive economy, *Econometrica*, 22, 265–90.

Arrow, K. and Honkapohja, S. (eds.) [1985], *Frontiers of Economics* (Oxford: Blackwell).

Arthur, W.B. [1989] Competing technologies, increasing returns, and lock-in by historical events, *Economic Journal*, 99, 116–31.

Arthur, W.B. [1990] Positive feedbacks in the economy, *Scientific American*, 262, 92–99.

Arthur, W.B. [1994] Inductive reasoning and bounded rationality, *American Economic Review: Papers and Proceedings*, 84, 406–11.

Arthur, W.B. [2014] *Complexity and the Economy* (Oxford: Oxford University Press).

Backhouse, R. [2004] History and equilibrium: A partial defense of equilibrium economics, *Journal of Economic Methodology*, 11, 291–300.

Balasko, Y. [2007] Out-of-equilibrium price dynamics, *Economic Theory*, 33, 413–35

Barro, R. and Grossman, H. [1971] A general disequilibrium model of income and employment, *American Economic Review*, 61, 82–93.

Becker, C. [1932] *The Heavenly City of the Eighteenth-Century Philosopher* (New Haven: Yale University Press).

Becker, G.S. [1965] A theory of the allocation of time, *Economic Journal*, 75, 493–517.

Benassy, J-P. [1975] Neo-Keynesian disequilibrium theory in a monetary economy, *Review of Economic Studies*, 42, 503–23.

Bicchieri, C. [1993] *Rationality and Coordination* (Cambridge: Cambridge University Press).

Binmore, K. [2007] Rational decisions in large worlds, *Annales d'Économie et de Statistique, 86*, 25–41.

Binmore, K. [2011] Interpreting knowledge in the backward induction problem, *Episteme, 8*, 248–61.

Blume, L. and Easley, D. [1998] Rational expectations and rational learning, in Majumdar, M (ed.) *Organizations with Incomplete Information: Essays in Economic Analysis* (Cambridge, UK: Cambridge University Press), 61–109.

Blume, L. and Easley, D. [2006] If you're so smart, why aren't you rich? Belief selection in complete and incomplete markets, *Econometrica, 74*, 929–66.

Boland, L. [1967] *Introduction to Price Theory*, unpublished manuscript.

Boland, L. [1971] An institutional theory of economic technology and change, *Philosophy of the Social Sciences, 1*, 253–58.

Boland, L. [1979a] A critique of Friedman's critics, *Journal of Economic Literature, 17*, 503–22.

Boland, L. [1979b] Knowledge and the role of institutions in economic theory, *Journal of Economic Issues, 8*, 957–72.

Boland, L. [1982] *The Foundations of Economic Method* (London: Geo. Allen & Unwin).

Boland, L. [1986] *Methodology for a New Microeconomics: The Critical Foundations* (Boston: Allen & Unwin).

Boland, L. [1989] *The Methodology of Economic Model Building: Methodology after Samuelson* (London: Routledge).

Boland, L. [1992] *The Principles of Economics: Some Lies My Teachers Told Me* (London: Routledge).

Boland, L. [1997] *Critical Economic Methodology: A Personal Odyssey* (London: Routledge).

Boland, L. [2003] *The Foundations of Economic Methodology: A Popperian Perspective* (London: Routledge).

Boland, L. [2010] Cartwright on 'Economics', *Philosophy of the Social Sciences, 40*, 530–8, doi: 10.1177/0048393109352867.

Boland, L. [2014] *Model Building in Economics: Its Purposes and Limitations* (New York: Cambridge University Press).

Bray, M. [1982] Learning, estimation and the stability of rational expectations equilibria, *Journal of Economic Theory, 26*, 318–39.

Brunnermeier, M. and Parker, J. [2005] Optimal expectations, *American Economic Review, 95*, 1092–118.

Bullard, J. and Suda, J. [2011] The stability of macroeconomic systems with Bayesian learners, Banque de France working paper no. 332.

Caldwell, B. [1980] Critique of Friedman's methodological instrumentalism, *Southern Economic Journal, 47*, 366–74.

Carlaw, K. and Lipsey, R. [2011] Sustained endogenous growth driven by structured and evolving general purpose technologies, *Journal of Evolutionary Economics, 21*, 563–93.

Carlaw, K. and Lipsey, R. [2012] Does history matter?: Empirical analysis of evolutionary versus stationary equilibrium views of the economy, *Journal of Evolutionary Economics, 22*, 735–66.

Carroll, L. [1871/1885] *Through the Looking Glass* (Boston: Lothrop Publishing Co.), 57–58.

Cassel, K.G. [1918/23] *The Theory of Social Economy* (New York: Harcourt).

Clower, R. [1955] Competition, monopoly and the theory of price, *Pakistan Economic Journal, 5*, 219–26.

Clower, R. [1959] Some theory of an ignorant monopolist, *Economic Journal, 69,* 705–16.

Clower, R. [1965] The Keynesian counterrevolution: A theoretical appraisal, in Hahn, F. and Brechling F. (eds.) *The Theory of Interest Rates* (London: Macmillan), 103–25.

Clower, R. [1994] Economics as an inductive science, *Southern Economics Journal, 60,* 805–14.

Coase, R. [1937] The nature of the firm, *Economica, 4 (NS),* 386–405.

Cross, J. [1973] A stochastic learning model of economic behavior, *Quarterly Journal of Economics, 87,* 239–66.

D'Agata, A. [2006] Entry and stationary equilibrium prices in a post-Keynesian growth model, in Salvadori, N. and Panico, C. (eds.), *Classical, Neo-Classical and Keynesian Views on Growth and Distribution,* (Cheltenham: Edward Elgar Pub.), 266–87.

David, P. [1985] Clio and the economics of QWERTY, *American Economic Review: Papers and Proceedings, 75,* 332–7.

Davidson, P. [1972] *Money and the Real World* (New York: Wiley).

Davidson, P. [1977/90] Money and general equilibrium, *Economie Appliquee,* reprinted in *Money and Employment: The Collected Writings of Paul Davidson, Volume 1* (London: Macmillan, 1990), 196–217.

Davidson, P. [1980] Causality in economics: A review, *Journal of Post-Keynesian Economics, 2,* 576–84.

Davidson, P. [1991] Is probability theory relevant for uncertainty? A post-Keynesian perspective, *Journal of Economic Perspective, 5,* 129–43.

Degler, C.N. [1991] *In Search of Human Nature: The Decline and Revival of Darwinism in American Social Thought* (Oxford: Oxford University Press).

De Vroey, M. [2002] Equilibrium and disequilibrium in Walrasian and neo-Walrasian economics, *Journal of the History of Economic Thought, 24,* 405–26.

Dore, M. [1984–85] On the concept of equilibrium, *Journal of Post-Keynesian Economics, 7,* 193–206.

Dorfman, R., Samuelson, P.A. and Solow, R.M. [1958] *Linear Programming and Economic Analysis* (New York: McGraw Hill).

Dosi, G. and Nelson, R. [1994] An introduction to evolutionary theories in economics, *Journal of Evolutionary Economics, 4,* 153–72.

Dotsey, M. and King, R.G. [1988] Rational expectations and the business cycle: A survey, *Federal Reserve Bank of Richmond Economic Review, 74,* 3–15.

Drazen, A. [1980] Recent developments in macroeconomic disequilibrium theory, *Econometrica, 48,* 283–306.

Driffill, J. [2011] The future of macroeconomics: Introductory remarks, *The Manchester School, 72,* 1–4.

Duarte, P.G. [2011] Recent developments in macroeconomics: The DSGE approach to business cycles in perspective, in Davis, J. and Hands, D.W. (eds.) *The Elgar Companion to Recent Economic Methodology* (Cheltenham and Northhampton: Edward Elgar), 375–403.

Düppe, T., and Weintraub, E.R. [2014] *Finding Equilibrium: Arrow, Debreu, McKenzie and the Problem of Scientific Credit* (Princeton, NJ: Princeton University Press).

Eaton, B.C., Eaton, D. and Allen, D. [2012] *Microeconomics: Theory and Applications, 8th Edition* (Toronto: Prentice Hall).

Evans, G. and Honkapohja, S. [2001] *Learning and Expectations in Macroeconomics* (Princeton, NJ: Princeton University Press).

Evans, G. and Honkapohja, S. [2009] Learning and macroeconomics, *Annual Review of Economics, 1,* 421–51.

Ezekiel, M. [1938] The Cobweb theorem, *Quarterly Journal of Economics, 52,* 255–80.

Finger, J.M. [1971] Is equilibrium an operational concept? *Economic Journal, 81*, 609–12.

Fisher, F. [1976] The stability of general equilibrium: Results and problems, in Artis, M.J. and Nobay, A.R. (eds.), *Essays in Economic Analysis: The Proceedings of the Association of University Teachers of Economics, Sheffield 1975* (Cambridge: Cambridge University Press), 3–29.

Fisher, F. [1981] Stability, disequilibrium awareness, and the perception of new opportunities, *Econometrica, 49*, 279–317.

Fisher, F. [1983] *Disequilibrium Foundations of Equilibrium Economics* (Cambridge: Cambridge University Press).

Fisher, F. [2003] Disequilibrium and stability, in Petri, F. and Hahn, F. [2003], 74–94.

Foster, J. [1997] The analytical foundations of evolutionary economics: From biological analogy to economic self-organization, *Structural Change and Economics Dynamics, 8*, 427–51.

Friedman, B. [1979] Optimal expectations and the extreme information assumptions of 'Rational Expectations' macromodels, *Journal of Monetary Economics, 5*, 23–41.

Friedman M. [1953] Methodology of positive economics, in *Essays in Positive Economics* (Chicago: University of Chicago Press), 3–43.

Frisch, R. [1936] On the notion of equilibrium and disequilibrium, *Review of Economic Studies, 3*, 100–105

Gabszewicz, J.J. [1985] Comment, in Arrow, K.J and Honkapohja, S. [1985], 150–69.

Gale, D. [2008] Money and general equilibrium, in Durlauf, S.N. and Blume, L.E. (eds.) *The New Palgrave Dictionary of Economics, 2nd Edition* (London: Palgrave Macmillan), doi:10.1057/9780230226203.1128.

Geanakoplos, J. [1987] Arrow-Debreu model of general equilibrium, in Eatwell, J., Milgate, M. and Newman, P. (eds.) *The New Palgrave: A Dictionary of Economics, 1* (London: The Macmillan Press), 116–24.

Gilboa, I. and Schmeidler, D. [2003] Inductive inference: An axiomatic approach, *Econometrica*, 1–26.

Gordon, D. and Hynes, A. [1970] On the theory of price dynamics, in Phelps, E. (ed.) *Microeconomic Foundations of Employment and Inflation Theory* (New York: Norton), 369–93.

Gordon, R. [1981] Output fluctuation and gradual price adjustment, *Journal of Economic Literature, 19*, 493–530.

Grossman, S. [1981] An introduction to the theory of rational expectations under asymmetric information, *Review of Economic Studies, 48*, 541–60.

Guesnerie, R. [2002]. Anchoring economic predictions in common knowledge, *Econometrica, 70*, 439–80.

Hahn, F. [1960] On the stability of competitive equilibrium, Working Paper No. 6, Bureau of Business and Economic Research, University of California at Berkeley, April, 1960.

Hahn, F. [1965] On some problems of proving the existence of an equilibrium in a monetary economy, in Hahn, F. and Brechling, F.P.R. (eds.) *The Theory of Interest Rates: Proceedings of a Conference Held by the International Economic Association* (London: Macmillan & Co), 126–35.

Hahn, F. [1970] Some adjustment problems, *Econometrica, 38*, 1–17.

Hahn, F. [1973] *On the Notion of Equilibrium in Economics* (Cambridge: Cambridge University Press).

Hahn, F. [1980] General equilibrium theory, *The Public Interest, Special Issue*, 123–38.

Hahn, F. [1981] General equilibrium theory, in Bell, D. and Kristol, I. (eds.) *The Crisis in Economic Theory* (New York: Basic Books), 123–38.

Hahn, F. and Negishi, T. [1962] A theorem on non-tâtonnement stability, *Econometrica*, 30, 463–69.

Hall, R. and Hitch, C. [1939] Price theory and business behaviour, *Oxford Economic Papers*, 2, 12–45.

Hands, D.W. [1996] Karl Popper on the myth of the framework: Lukewarm Popperians +1, unrepentant Popperians –1, *Journal of Economic Methodology*, 3, 317–22.

Hansen, L.P. and Sargent, T. [1980] Formulation and estimating dynamic linear rational expectations models, *Journal of Economic Dynamics and Control*, 2, 7–46.

Hart, O. [1975] On the optimality of equilibrium when the market structure is incomplete, *Journal of Economic Theory*, 11, 418–33.

Hart, O. [1985] Imperfect competition in general equilibrium: An overview of recent work, in Arrow, K. and Honkapohja, S. [1985], 100–49.

Hart, S. and Mas-Colell, A. [2003] Uncoupled dynamics do not lead to Nash equilibrium, *American Economic Review*, 93, 1830–36.

Hayek, F. [1933/39] Price expectations, monetary disturbances and malinvestments, in *Profits, Interest and Investments* (London: Routledge), 350–65. Published form of a lecture delivered on December 7, 1933, in the *Sozialökonomisk Samfund* in Copenhagen.

Hayek, F. [1936] The mythology of capital, *Quarterly Journal of Economics*, 50, 199–228.

Hayek, F. [1937] Economics and knowledge, *Economica*, 4 (NS), 33–54.

Hayek, F. [1945] The uses of knowledge in society, *American Economic Review*, 35, 519–30.

Heckman, J. [2003] Conditioning, causality and policy analysis, *Journal of Econometrics*, 112, 73–8.

Henry, J. [1983–84] On equilibrium, *Journal of Post-Keynesian Economics*, 6, 214–29.

Hicks, J. [1939/46] *Value and Capital: An Inquiry into Some Fundamental Principles of Economic Theory, 2nd Edition* (Oxford: Clarendon Press).

Hicks, J. [1956] *A Revision of Demand Theory* (Oxford: Clarendon Press).

Hicks, J. [1976] Some questions of time in economics, in Tang, A., Westfield, F. and Worley, J. (eds.) *Evolution, Welfare and Time in Economics* (Toronto: Heath), 135–51.

Hicks, J. [1979] *Causality in Economics* (Oxford: Basil Backwell).

Hicks, J. and Allen, R. [1934] A reconsideration of the theory of value, *Economica, 1* (NS), 54–76 and 196–219.

Hirsch, A. and de Marchi, N. [1984] Methodology: A comment on Frazer and Boland, *American Economic Review*, 74, 782–88.

Hitchcock, A. [1955] Television monologue.

Hodgson, G. [1997] The evolutionary and non-Darwinian economics of Joseph Schumpeter, *Journal of Evolutionary Economics*, 7, 131–45.

Hodgson, G. and Knudsen, T. [2006] Why we need a generalized Darwinism, and why generalized Darwinism is not enough, *Journal of Economic Behavior & Organization*, 61, 1–19.

Holland, J., Holyoak, J., Nisbett, R. and Thagard, P. [1986] *Induction: Processes of Inference, Learning and Discovery* (Cambridge: MIT Press).

Hoover, K. [1984] Methodology: A comment on Frazer and Boland, *American Economic Review*, 74, 789–92.

Hoover, K. [2001] *The Methodology of Empirical Macroeconomics* (Cambridge: Cambridge University Press).

Hoover, K. [2012] Microfoundational programs, in Duarte, P.G. and Lima, G.T. (eds.) *Microfoundations Reconsidered: The Relationship of Micro and Macroeconomics in Historical Perspective*, (Cheltenham: Edward Elgar), 29–61.

Jarvie, I. [1972] *Concepts and Society* (London: Routledge & Kagen Paul).

Kahneman, D. and Tversky, A. [1979] Prospect theory: An analysis of choice under risk, *Econometrica*, 47, 263–91.

Kaldor, N. [1934] The equilibrium of the firm, *Economic Journal*, 44, 60–76.

Kaldor, N. [1983] Keynesian economics after fifty years, in Worswick and Trevithick [1983], 1–28.

Keynes, J.M. [1921] *A Treatise on Probability* (London: MacMillian and Co.).

Keynes, J.M. [1936] *General Theory of Employment, Interest and Money* (New York: Harcourt, Brace and World).

Keynes, J.M. [1937] The general theory of employment, *Quarterly Journal of Economics*, 51, 209–23.

Kirman, A. [1992] Whom or what does the representative individual represent?, *Journal of Economic Perspectives*, 6, 117–36.

Kirman, A. [2003] General equilibrium: Problems, prospects and alternatives: An attempt at synthesis, in Petri F. and Hahn F. [1983], 468–520.

Kirman, A. [2011] Walras' unfortunate legacy, in Bridel. P. (ed.), *General Equilibrium Analysis: A Century after Walras* (New York: Routledge), 109–33.

Knudsen, T. [2004] General selection theory and economic evolution: The price equation and the replicator/interactor distinction, *Journal of Economic Methodology*, 11, 147–73.

Koopmans, T. [1957] *Three Essays on the State of Economic Science* (New York: McGraw-Hill).

Kreps, D. [1990] *Game Theory and Economic Modelling* (New York: Oxford University Press).

Lachmann, L.M. [1976] From Mises to Shackle: An essay on Austrian economics and the kaleidic society, *Journal of Economic Literature*, 14, 54–62.

Lachmann, L.M. [1982] The salvage of ideas: Problems of the revival of Austrian economic thought, *Journal of Institutional and Theoretical Economics*, 138, 629–45.

Lancaster, K. [1966] A new approach to consumer theory, *Journal of Political Economy*, 74, 132–57.

Lancaster, T. [2004] *An Introduction to Modern Bayesian Econometrics* (Oxford: Blackwell Publishing).

Latsis, S. [1972] Situational determinism in economics, *British Journal for the Philosophy of Science*, 23, 207–45.

Leamer, E. [1983] Let's take the con out of econometrics, *American Economic Review*, 73, 31–43.

Leijonhufvud, A. [1983] What would Keynes have thought about rational expectations?, in Worswick, G.D.N. and Trevithick, J. [1983], 179–205.

Leijonhufvud, A. [1993] Towards a not-too-rational macroeconomics, *Southern Economics Journal*, 60, 1–13.

Lester, R. [1946] Shortcomings of marginal analysis for wage-employment problems, *American Economic Review*, 36, 63–82.

Lichtenstein, S. and Slovic, P. [1971] Reversal of preferences between bids and choices in gambling decision, *Journal of Experimental Psychology*, 89, 46–55.

Loasby, B. [1976] *Choice, Complexity and Ignorance* (Cambridge: Cambridge University Press).

Long, J. and Plosser, C. [1983] Real business cycles, *Journal of Political Economy*, *91*, 39–69.

Lucas, R. [1976] Econometric policy evaluation: A critique, in the *Carnegie-Rochester Conference Series on Public Policy*, *1*, 19–46.

Lucas, R. and Prescott, E. [1971] Investment under uncertainty, *Econometrica*, *39*, 659–81.

Machlup, F. [1958] Equilibrium and disequilibrium: Misplaced concreteness and disguised politics, *Economic Journal*, *68*, 1–24.

Makridakis, S. and Hibon, M. [1979] Accuracy of forecasting: An empirical investigation, *Journal of the Royal Statistical Society Series A*, *142*, 97–145.

Makridakis, S. and Hibon, M. [2000] The M3-competition: Results, conclusions and implications, *International Journal of Forecasting*, *116*, 451–76.

Mariotti, M. [1995] Is Bayesian rationality compatible with strategic rationality?, *Economic Journal*, *105*, 1099–109.

Marshall, A. [1920/64] *Principles of Economics, 8th Edition* (London: Macmillan).

McKenzie, L.W. [1954] On equilibrium in Graham's model of world trade and other competitive systems, *Econometrica*, *22*, 147–61.

McKenzie, L.W. [1987] General equilibrium, in Eatwell, J., Milgate, M. and Newman, P. (eds.) *The New Palgrave: A Dictionary of Economics, 2* (London: The Macmillan Press), 498–512.

Milgate, M. [1987] Equilibrium: Development of the concept, in Eatwell, J., Milgate, M. and Newman, P. (eds.) *The New Palgrave: A Dictionary of Economics, 2* (London: The Macmillan Press), 179–83.

Mitra-Khan, B.H. [2008] Debunking the myths of computable general equilibrium models, working paper 2008-1, *Schwartz Center for Economic Policy Analysis, New School for Social Research*.

Muth, J. [1961] Rational expectations and the theory of price movements, *Econometrica*, *29*, 315–35.

Muth, J. [1987] Discussion of Schips' paper, in Wold, H. (ed.) *Theoretical Empiricism: A General Rationale for Scientific Model-Building* (New York: Paragon House), 97–102.

Muth, J. [1994] Does economics need theories?, in Klein, P. (ed.) *The Role of Economic Theory* (Boston: Kluwer Academic Press), 97–119.

Nachbar, J. [2001] Bayesian learning in repeated games of incomplete information, *Social Choice and Welfare*, *18*, 303–26.

Negishi, T. [1960–1] Monopolistic competition and general equilibrium, *Review of Economic Studies*, *28*, 196–201.

Negishi, T. [1985] *Economic Theories in a Non-Walrasian Tradition* (Cambridge: Cambridge University Press).

Nelson, R. [2001] Evolutionary economics – The state of the science, presented at New Perspectives on Telecommunications and Pharmaceuticals in Europe and the United States, Institute for Applied Economics and the Study of Business Enterprise Conference on Evolutionary Economics, Johns Hopkins University, March 30–1.

Nelson, R. [2013] Demand, supply, and their interaction on markets, as seen from the perspective of evolutionary economic theory, *Journal of Evolutionary Economics*, *23*, 17–38.

Nelson, R. and Winter, S. [1974] Neoclassical vs. evolutionary theories of economic growth: Critique and prospectus, *Economic Journal*, *84*, 886–905.

Nelson, R. and Winter, S. [1982] *An Evolutionary Theory of Economic Change* (Cambridge: Harvard University Press).

Nelson, R. and Winter, S. [2002] Evolutionary theorizing in economics, *Journal of Economic Perspectives*, *16*, 23–46.

Neumann, J. von [1937/45] A model of general equilibrium, *Review of Economic Studies*, *13*, 1–9. Published version of a paper that was first read in 1932 at Princeton University.

Nikaido, H. [1960/70] *Introduction to Sets and Mappings in Modern Economics* (Amsterdam: North Holland).

North, D. [1978] Structure of performance: The task of economic history, *Journal of Economic Literature*, *36*, 963–78.

Ostroy, J. [1987] Money and general equilibrium, in Eatwell, J., Milgate, M. and Newman, P. (eds.) *The New Palgrave: A Dictionary of Economics, 3* (London: The Macmillan Press), 515–18.

Pashigian, P. [1987] Cobweb theorem, in Eatwell, J., Milgate, M. and Newman, P. (eds.) *The New Palgrave: A Dictionary of Economics, 1* (London: The Macmillan Press), 463–64.

Pesaran, M.H. and Smith, R. [2011] Beyond the DSGE straitjacket, *The Manchester School*, 5–38.

Petri, F. and Hahn, F. [2003] *General Equilibrium: Problems and Prospects* (London: Routledge).

Pindyck, R. and Rubinfeld, D. [1998] *Econometric Models and Economic Forecasts* (Boston: McGraw-Hill).

Poirier, D. [1988] Frequentist and subjectivist perspectives on the problem of model building in economics, *Journal of Economic Perspectives*, *2*, 121–70.

Popper, K. [1963] *Conjectures and Refutations* (New York: Basic Books).

Popper, K. [1966] *The Open Society and Its Enemies, Volume II* (New York: Harper & Row).

Price, G.R. [1972] Extension of covariance selection mathematics, *Annals of Human Genetics*, *35*, 485–90.

Price, G.R. [1995] The nature of selection, *Journal of Theoretical Biology*, *175*, 389–96.

Richardson, G. [1959] Equilibrium, expectations and information, *Economic Journal*, *69*, 225–37.

Robinson, J. [1933/65] *The Economics of Imperfect Competition* (London: MacMillan).

Robinson, J. [1953–54] Production function and the theory of capital, *Review of Economic Studies*, *21*, 81–106.

Robinson, J. [1962] Book review, *Economic Journal*, *72*, 690–92.

Robinson, J. [1974] History versus equilibrium, *Thames Papers in Political Economy* (London: Thames Polytechnic).

Robinson, J. [1978] *Contributions to Modern Economics* (Oxford: Blackwell).

Robson, A. [2002] Evolution and human nature, *Journal of Economic Perspectives*, *16*, 89–106.

Russell, B. [1945] *A History of Western Philosophy,* (New York: Simon & Schuster).

Rutherford, M. [2001] Institutional economics: Then and now, *Journal of Economic Perspectives*, *15*, 173–94.

Samuels, W. [1997] On the nature and utility of the concept of equilibrium, *Journal of Post-Keynesian Economics*, *20*, 77–88.

Samuelson, P. [1947/65] *Foundations of Economic Analysis* (New York: Atheneum) Published version of PhD thesis.

Sargent. T. [1973] Rational expectations, the real rate of interest and the natural race of unemployment, *Brookings Papers on Economic Activity*, *2*, 429–72.

Sargent. T. [1993] *Bounded Rationality in Macroeconomics* (Oxford: Oxford University Press).

Sargent, T. [2008] Evolution and intelligent design, *American Economic Review*, 98, 5–37.

Scarf, H. [1973] *The Computation of Economic Equilibria* (with the collaboration of T. Hansen), (New Haven, CT: Yale University Press).

Scarf, H.E. [1967] The approximation of fixed points of a continuous mapping, *SIAM Journal on Applied Mathematics*, 15, 1328–43.

Schumpeter, J. [1909] On the concept of social value, *Quarterly Journal of Economics*, 23, 213–32.

Schumpeter, J. [1954] *History of Economic Analysis* (New York: Oxford University Press).

Scitovsky, T. [1976] *Joyless Economy* (Oxford: Oxford University Press).

Sent, E.-M. [1997] Sargent versus Simon: Bounded rationality unbound, *Cambridge Journal of Economics*, 21, 323–38.

Sent, E.-M. [2002] How (not) to influence people: The contrary tale of John F. Muth, *History of Political Economy*, 32, 291–319.

Shackle, G. [1972] *Epistemics and Economics* (Cambridge: Cambridge University Press).

Simon, H. [1953] Causal ordering and identifiability, in Hood, W. and Koopmans, T. (eds.), *Studies in Econometric Method* (New York: Wiley), 49–74.

Simon, H. [1955] A behavioral model of rational choice, *Quarterly Journal of Economics*, 69, 99–118.

Simon, H. [1957] *Models of Man, Social and Rational: Mathematical Essays on Rational Human Behavior in a Social Setting* (New York: Wiley)

Simon, H. [1979] Rational decision making in business organizations, *American Economic Review*, 69, 493–513.

Smithies, A. [1942] Process analysis and equilibrium analysis, *Econometrica*, 10, 26–38.

Solow, R. [1979] Alternative approaches to macroeconomic theory: A partial view, *Canadian Journal of Economics*, 12, 339–54.

Solow, R. [2008] The state of macroeconomics, *Journal of Economic Perspectives*, 22, 243–49.

Spear, S. [1989] Learning rational expectations under computability constraints, *Econometrica*, 57, 889–910.

Spencer, H. [1864] *The Principles of Biology* (London: Williams and Norgate).

Sraffa, P. [1926] The laws of returns under competitive conditions, *Economic Journal*, 38, 535–50.

Sraffa, P. [1960] *Production of Commodities by Means of Commodities* (Cambridge: Cambridge University Press).

Stigler, G. [1961] The economics of information, *Journal of Political Economy*, 69, 213–25.

Stigler, G. and Becker, G. [1977] De gustibus non est disputandum, *American Economics Review*, 67, 76–90.

Stiglitz, J. [1975] Information and economic analysis, in Parkin, M. and Nobay, A. *Current Economic Problems: The Proceedings of the [Annual Mmeeting of the] Association of University Teachers of Economics, Manchester, 1974* (Cambridge: Cambridge University Press), 27–52.

Sugden, R. [1998] The role of inductive reasoning in the evolution of conventions, *Law and Philosophy*, 17, 377–410.

*The American Heritage Dictionary of the English Language, 4th Edition* [2000/09] (New York: Houghton Mifflin Company).

Uzawa, H. [1962] On the stability of Edgeworth's barter process, *International Economic Review*, *3*, 218–32.

Vives, X. [1993] How fast do rational agents learn? *Review of Economic Studies*, *60*, 329–47.

Vriend, N. [2002] Was Hayek an Ace? *Southern Economics Journal*, *69*, 811–41.

Vromen, J. [2006] Routines, genes and program-based behavior, *Journal of Evolutionary Economics*, *16*, 543–60.

Vromen, J. [2012] Ontological issues in evolutionary economics: The debate between Generalized Darwinism and the Continuity Hypothesis, in Mäki, U. (ed.) *Elsevier Handbook of the Philosophy of Science, Volume 13: Philosophy of Economics* (Amsterdam: North Holland), 737–63.

Wald, A. [1936/51] On some systems of equations of mathematical economics, *Econometrica*, *19*, 368–403, English version of 'Über einige Gleichungssysteme der mathematischen Ökonomie', *Zeitschrift fibr Nationalökonomie*, *7*, 1936, 637–70.

Waldrop, M. [1992] *Complexity: The Emerging Science at the Edge of Order and Chaos* (New York: Simon & Schuster).

Walras, L. [1874] *Éléments d'économie politique pure, ou théorie de la richesse social* (Lausanne: L. Corbaz).

Weintraub, E.R. [1985a] *General Equilibrium Analysis: Studies in Appraisal* (Cambridge, UK: Cambridge University Press).

Weintraub, E.R. [1985b] Joan Robinson's critique of equilibrium: An appraisal, *American Economic Review: Papers and Proceedings*, *75*, 146–69.

Weintraub, E.R. [2005] On Lawson on equilibrium, *Journal of Post-Keynesian Economics*, *27*, 445–54.

Weintraub, E.R. [2011] Retrospectives: Lionel W. McKenzie and the proof of the existence of a competitive equilibrium, *Journal of Economic Perspectives*, *25*, 199–215.

Wible, J. [1982] Friedman's positive economics and philosophy of science, *Southern Economic Journal*, *49*, 50–60.

Williamson, O. [1985] *The Economic Institutions of Capitalism: Firms, Markets, Relational Contracting* (London: Collier Macmillan Publishers).

Williamson, O. [2000] The new institutional economics: Taking stock, looking ahead, *Journal of Economic Literature*, *38*, 595–613.

Winter, S. [1964] Economics 'natural selection' and the theory of the firm, *Yale Economic Essays*, *4*, 225–72.

Winter, S. [1971] Satisficing, selection and the innovating remnant, *Quarterly Journal of Economics*, *85*, 237–61.

Winter, S. [2014] The future of evolutionary economics: Can we break out of the beach-head?, *Journal of Institutional Economics*, *11*, 613–44.

Witt, U. [2008] Evolutionary economics, Durlauf, S.N. and Blume, L.E. (eds.) *The New Palgrave Dictionary of Economics, 2nd Edition* (London: Palgrave Macmillan), doi:10.1057/9780230226203.0511.

Witt, U. [2014] The future of evolutionary economics: Why the modalities of explanation matter, *Journal of Institutional Economics*, *11*, 645–64.

Woodford, M. [2013] Macroeconomic analysis without the Rational Expectations Hypothesis, *Annual Review of Economics*, *5*, 303–46.

Worswick, D. and Trevithick, J. (eds.) [1983] *Keynes and the Modern World* (Cambridge: Cambridge University Press).

Zame, W. [2008] General equilibrium (new developments), Durlauf, S.N. and Blume, L.E. (eds.) *The New Palgrave Dictionary of Economics, 2nd Edition* (London: Palgrave Macmillan), doi:10.1057/9780230226203.0624.

# NAMES INDEX

# SUBJECT INDEX

Austrian economists, 80
average cost, 32, 44–45, 68, 111, 133,
    148–49, 180, 192, 198

banking, 142
behavioral sciences, 168
behavioural economics, 156,
    229–30, 233
behavioural finance, 231
behavioural microeconomics, 157
  model of firm, 157
biological growth rate, 93–94
business cycle theory, 141

causality, 17, 68, 70–71, 73–75,
    92, 94–95
  cause vs. effect, 70
collusion, 3, 59, 91, 99, 103–4
  absence of, 91
  explicit, 55
  implicit, 55
comparative advantage, 52
complexity, 159–62, 229, 231–32
  computational, 230
complexity economics, 159, 161–62,
    229, 231–32
  case-by-case, 233
  computer program model, 38, 91,
    161–63, 169
  path dependency, 154–55, 163,
    229, 233–34
  sociological interactions, 161
conjectural theory, false, 85
conjecture, theoretical, 85
constraints, non-natural, 5
consumer, theory of, 88, 194
  budget line, 87, 114, 213–14

conjectural indifference map, 88, 213,
    216–17, 223
convexity assumption, 87, 131
demand curve, 1–3, 8, 26, 28,
    32–34, 36–45, 50, 59, 85,
    101, 103–5, 148, 165, 174,
    180–84, 192, 210, 216,
    219–20, 223
  Giffen effect, 28, 219–20
demand elasticity, 33, 181–82
exogenously diverse tastes, 226
expected utility, 6, 8, 22, 30, 48, 51,
    81, 83, 85, 87, 89, 100, 104,
    115, 120, 125, 141, 165, 175,
    177, 184, 186–87, 193–94,
    197, 207, 209, 213–14,
    216–17, 220, 230
indifference curve, 88–89, 112
  slope, 214
  strictly convex, 87, 89, 214
  map, 89
indifference map, 87–89, 112, 214,
    216–17, 214, 216–17, 223
  insufficient knowledge of, 213
learning consumer, 209
luxury goods, 219
preference map, 83, 85, 87–89, 113,
    166, 186, 207, 213–14
  strictly convex, 87, 89, 214
preferences, 50–51, 53, 84–85, 89, 94,
    108, 132, 139, 147, 156, 164,
    209, 230
  psychologically-given, 94
  sociological aspects, 157
  tastes, 4, 21, 93–94, 97–98, 157,
    215, 217, 222, 227
satisficing, 81, 149